CRYSTAL CLEAR

CRYSTAL CLEAR

The Struggle for Reliable Communications Technology in World War II

RICHARD J. THOMPSON, JR.

IEEE PRESS

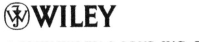

A JOHN WILEY & SONS, INC., PUBLICATION

Published by John Wiley & Sons, Inc., Hoboken, New Jersey.
Published simultaneously in Canada.

For general information on our other products and services please contact our Customer Care Department within the United States at (800) 762-2974, outside the United States at (317) 572-3993 or fax (317) 572-4002.

Wiley also publishes its books in a variety of electronic formats. Some content that appears in print, however, may not be available in electronic formats. For more information about Wiley products, visit our web site at www.wiley.com.

Library of Congress Cataloging-in-Publication Data is available.

ISBN 978-1-118-10464-4 (paper edition)

10 9 8 7 6 5 4

CONTENTS

ACKNOWLEDGMENTS

This book would have never come into being without the help of a great many people. Among those deserving special praise are Donald Frazier, for helping me to develop my "recessive history gene"; Carlene Stephens, for her patient reading of and extremely helpful comments on the manuscript and book prospectus; Ken Burch, for the wealth of documents and photos that he provided me during the writing of the manuscript; Gary Wilson, for being a great friend and supporting me in just about everything I ever attempted; John Vig, for putting me in touch with very helpful people in the field; Richard Bingham and Chris McGahey, for providing materials from the Fort Monmouth archives; Chuck Bliley, for information on his family's company; Phil Craik, for his help in working with archival film footage; Michael Gannon, for helping me search for sunken quartz ships; Robert Schultz, for his inside information on crystal grinding teams; Canon Bradley, for his insider's view of the crystal industry and his reading of the manuscript; and Joel Kleinman (*QST Magazine*) and Jack Kusters (IEEE), for helping me get my first pieces on the wartime crystal industry into print.

I am very grateful to the following persons and organizations for funding and other support during the research and writing phases of this project: Roger and Kim Ward (Ward-Bottom Fund), McMurry University, the Sam Taylor Foundation, and the McWhiney Foundation. Lastly, I am very appreciative of all the help and encouragement that I received from archivists at the following institutions: the Center of Military History (particularly Bob Wright and David Hogan), the National Archives II, the Smithsonian National Museum of American History, the Library of Congress, the Motorola Archives, and Carlisle Barracks.

To my family, I am indebted for a lifetime of love and support. My wife, Laura, supported me every step of the way, even serving as research assistant on two very successful trips to the Archives; for all that she has done I owe her more than I can ever hope to repay.

Lastly, a special word of thanks is owed to Dr. Virgil Bottom. It was a reading of his memoirs and many long conversations with him that lead me to pursue this project. In my personal and professional life (both as historian and physicist), Dr. Bottom served as an outstanding role model. It is to him and all the other men and women of this story that this book is dedicated. I hope I've done them justice.

ALBANY, NY

INTRODUCTION: *"WE WERE HEAVILY ARMED, AND WE HAD CRYSTALS"*

Private First Class Irwin Gottlieb joined his unit in France on June 9, 1944. Trained as a machine gunner, he was assigned to the First Reconnaissance Cavalry Troop of the First Infantry Division. From D+3 till VE Day, Gottlieb and his unit saw almost continuous action, often miles ahead of the rest of the division. It was not uncommon for German troops to avoid contact with the 30-man troop, possibly confusing them for the head of a much larger armored column or, at the very least, hoping to prevent disclosing their position to the recon unit. Firefights did erupt, however; Gottlieb himself was wounded during the last days of the war. In later life, when asked how his unit defended itself against often-times much larger German units, Gottlieb would invariably reply "we were heavily armed, and we had crystals."[1]

Being "heavily armed" is easy to understand: with an assortment of .30- and .50-caliber machine guns, 37-mm cannon, and 60-mm mortars, they could give as good as they got. The meaning of "and we had crystals" is not quite as obvious. What he meant by this was that his unit possessed quartz crystal-controlled radio equipment. The saga of how such equipment came to be regarded by a front-line combat veteran as a weapon as vital to survival as machine guns and mortars is the object of this work.

What made their radio equipment such a powerful weapon were the quartz crystal units that controlled their transmitting and receiving frequencies. Comprised of little more than a fingernail-sized wafer of quartz, these crystal units provided the operating stability that ultimately allowed instant and

1

dependable radio communications to be taken for granted by the men in the fields, the tanks, and the airplanes.

Though the outward appearance of the quartz crystal unit suggested a very simple device, their manufacture required methods of exacting standards and extreme precision. So much so that prior to WWII they were produced one at a time, by hand, in a small number of companies across the country. The entire output of the crystal "industry" in 1941 was only 100,000 units. However, by the end of the war, a full-fledged industry numbering nearly 150 manufacturers was turning out over two million units *per month*.

Quartz crystals went from a 19th-century scientific curiosity to the focus of a massive military and industrial program during the Second World War. The largely untold story of this transformation is one of science and technology and the problems of peace-time military planning. It deals with the conflict between the established arms of the military and the rapidly evolving and expanding ones born of the First World War. It involves unprecedented cooperation among and between various government agencies, independent branches of the military, and private industry in order to design, build, supply, and support a wartime mass production industry where none had existed prior to the attack on Pearl Harbor.

In this book, the story of the quartz crystal industry is divided into five sections. The initial section serves as an introduction; covering the history and the science of the piezoelectric effect of quartz, its use in radio electronics, and the ultimate acceptance of this mode of frequency control by the U. S. Army Signal Corps, the branch of the Army with overall responsibility for communications and its corresponding technologies. Sections 2, 3, and 4 address what I refer to as the "Three Crises." The first is the crisis brought on by America's abrupt entry into the Second World War. Suddenly faced with enormous needs for reliable military communication equipment, the country had no mass production industry to produce the crystal units needed for frequency control. Worse still, there existed no mass production techniques or equipment to be utilized even if an industry could somehow have been conjured up immediately from thin air. Being the primary agency involved with the development of new communications technologies, the solution to this problem fell to the U.S. Army's Office of the Chief Signal Officer.

The second crisis involves the problem of supplying the crystal industry with the unprecedented amounts of raw quartz needed for manufacturing the crystal units. The only sources of "radio grade" quartz available during the war were within the interior regions of Brazil. Defined by Congress to be a "strategic and critical" material, the problems of how to increase the production, purchase, and transport of quartz to the United States were faced primarily by the departments of the Executive Branch of the civilian government. The third crisis came about after it appeared that the first two had been overcome by the Signal Corps and the government. It had to do with something known as the "Aging Problem," an inevitable failure of the crystal units due to the manner in which they were manufactured. Representing a communications research and development challenge, the solution of this problem came primarily from those most experienced in this field, the scientists and engineers of the Signal Corps laboratories.

The final section of the book attempts to put the story of the quartz crystal unit into context with much better-known industrial and scientific contributions to the war effort. In terms of rapid industrial growth and dramatic increases in output, this story is not unique. A great many industries (such as aircraft and ship manufacturers) grew in size during the war. A great many other sectors of industry (such as the automotive industry) retooled their plants for the production of war materiel. Completely new industries (particularly the synthetic rubber industry) were created in this country by scientists, engineers, and industrialists where none had existed before. What is truly unique about the crystal industry is that it was *invented from scratch*. There was no mass production industry to expand. To enter the crystal business took much more than a simple "retooling" of manufacturing plants; in early 1942, no one really knew *how* to mass produce crystal units. Even the synthetic rubber industry, essentially a new entity in the United States, was built along the lines of existing programs in other countries. No such blueprints existed for the quartz crystal industry.

The complete story of this wartime effort has never been told in any unified way. Though a handful of reports and conference presentations have been produced that recount the history of particular groups involved with the quartz crystal industry, this book, based on hundreds of primary documents, correspondence, and interviews, is the first to attempt to portray the entire enterprise.

This story more than anything else is one of invention. At its heart, this book is about the quartz crystal oscillator, a product of pure research that was almost instantaneously embraced by the amateur radio community. However, the invention theme encompasses much more of the story. It also involves the inventiveness of the early pioneers of the crystal industry, developing the tools and techniques needed to manufacture the crystal units. It includes the work of the Signal Corps and the U.S. government to invent a mass production industry for an item whose crucial importance to the military was never fully realized until the war began. The new methods of business cooperation and the ways of confronting the age-old problems of supply and demand that were developed with respect to the crystal industry can also be considered inventions. Overall, this is a story of an interconnected web of inventors (scientists, industrialists, basement hobbyists, and military administrators) and inventions (material objects, techniques, and ideas). The overall success of this wartime program can quite possibly be linked directly to the fact that it essentially had no history; no previous modes of thought and action that could inhibit the free-thinking and inventiveness on the part of the participants.

This book aims to show that the crystal program played just as important a role as radar or the atomic bomb in terms of its scientific and engineering contributions to the war effort and as any other sector of industry in terms of its rapid response to the challenges brought on by the war. Furthermore, the development of a mass production industry for quartz oscillators during WWII had far-reaching effects on late 20th-century technology and society. Today, nearly everything that requires some type of timing or frequency control depends on a quartz oscillator. This includes cell phones, color television, computers, watches and clocks, wire-based multisignal telephone technology, and many other items upon which our modern society depends. In fact, it can be argued that the move toward crystal control, with its reliance on a truly 20th-century, solid-state technology, presaged society's coming dependence upon the transistor and the integrated circuit and marks the very beginning of the evolution from an analog to a digital world.

"Pass not the shapeless lump of crystal by,
Nor view the icy mass with careless eye:
All royal pomp its value far exceeds.

And all the pearls the Red Sea's bosom breeds,
This rough and uniform'd stone, without a grace,
Midst rarest treasure holds the chiefest place."
Claudius, 14 AD[2]

"We have faith that future generations will know that here, in the middle of the twentieth century, there came a time when men of good will found a way to unite, and produce, and fight to destroy the forces of ignorance, and intolerance, and slavery, and war."
Franklin Delano Roosevelt

1

FROM WIRE TO WIRELESS: THE DEVELOPMENT—AND ACCEPTANCE—OF TACTICAL RADIO

Ultimately, the argument came down to one inescapable fact: you just couldn't run a telephone line to a tank or an airplane and have them do what they were designed to do. Like it or not, by 1940, tactical radio was a military necessity. This change in communications doctrine took place very slowly during the interwar years of the 1920s and 1930s and paralleled the developments of two military branches spawned by the previous war: the Air Corps and the Armored Forces. As the usefulness and importance of these two branches increased in the eyes of the nation's military leaders, so did the attention to their needs. Primary among these needs was dependable, *mobile*, communications technology—radio.

Radio saw very little use in World War I; wire, both telephone and telegraph, was the primary medium of communication, supplemented by runners, motorcycle messengers, and carrier pigeons. This arrangement still dominated Signal Corps planning on the eve of the Second World War. It is sometimes said that the military always prepares to fight the *previous war*, never the next one. In terms of the Signal Corps during the years between the World Wars, this seems very true. It is difficult, however, to really blame them. The primary charge of the Signal Corps was to develop the communications equipment needed for the next conflict and have it ready and available whenever that conflict should break out. Essentially, they were expected to predict the future, and such predictions are usually based on the past. In the previous war, the ground forces fought on nearly stationary fronts. *Mobility*, if it could be claimed at all, usually meant movements parallel to the front lines, not across. Thus, wire and cable were the perfect technology upon which to build a communications system stretching from the rear areas "all the way to the barbed wire."[1]

With a nod toward a possible increase in mobility, the Signal Corps of the 1920s and 1930s concentrated on ways of laying wires faster, on the development of cable and amplification methods for longer distance transmission, and on improved field telephones and switchboard equipment for front-line units. Radio, if it were to play any tactical role, would be seen as a "stand by" or emergency piece of equipment.[2] Again, it is hard to fault the Signal Corps for this mind set due to the many benefits of wire: it had a long established history, it had proven its usefulness in battle, there had been time for the Signal Corps to work out the problems that had surfaced during the previous war, and it was familiar—it did not take a lot of training for someone to be able to pick up a conventional telephone and talk into it. The U.S. military was by no means the only one to view radio in this manner. The 1935 edition of the British Army's *Field Service Regulations* included no mention of radio whatsoever and the 1934 declaration of Germany's chief tank tactician, Heinz Guderian, of his desire to lead his tank divisions "from the front, by wireless" was regarded as "nonsense!" by his superiors. Guderian's ideas were vindicated, however, in both the invasions of France (where only one in five French tanks was supplied with radios) and of Russia (where only battalion commanders' tanks were so equipped).[3]

The communications picture for the Air Corps was a lot clearer; as wire obviously was not an option, radio had to be utilized. Much of the early Signal Corps research and development work carried out with regards to radio was geared toward airborne sets. Evidence of the importance of this work is the creation in 1927 of the Signal Corps' Aircraft Radio Laboratory (ARL) at Wright Field in Dayton, Ohio, placed in charge of development and testing of new airborne radio equipment.[4] In some ways, even this was seen as a compromise on the part of the Signal Corps. While still considering radio of secondary interest, it did not want to lose complete control of its development to the rapidly expanding Air Corps. Though the Signal Corps prevailed in its dominance over radio research and development, its association with the Air Corps was less than comfortable (with the Air Corps accusing the Signal Corps of "slow and unimaginative" research and development, and the Signal Corps constantly on the lookout for challenges to its turf from the Air Corps).[5] The work continued in spite of these difficulties until, by 1936, it could be claimed that all Army war planes were fully equipped with radio technology.[6] In fact, a large proportion of the Signal Corps' 1937 appropriation of $5.6 million went toward the needs of airborne radio.[7]

The struggle for improved communications was much harder for the ground forces. As discussed above, wire-based communications were a proven concept for ground warfare and, in the opinion of the Signal Corps, only needed to be strengthened; radio did not figure prominently in the Signal Corps' plans for the next war.[8] Opposition to this point of view came primarily from the newest ground arm, the Mechanized Cavalry (later to be known as the Armored Forces). Armored tanks had been developed in the last years of the previous war for the express purpose of reintroducing cross-front mobility. Some way had to be found to break out of the trenches and move toward *and beyond* the enemy's positions; the tank had held this promise. During the 1930s, a new doctrine began to emerge that placed a greater emphasis on both mechanization and mobility, with the armored forces serving as spearheads of rapidly moving campaigns. This new doctrine could not be expressed within the Signal Corps communications model. Along with needing constant communications among the armored units and their support troops, a rapidly advancing column would surely outrun its wire-based communications (examples of this will be discussed in Chapter 3).[9]

A third military entity to play a role in the development of a tactical radio doctrine was that of the anti-aircraft batteries and the Air Warning System. With the increased speed of aircraft, it became essential that a method of early detection and warning of approaching aircraft be developed. In this regard, the Signal Corps played perhaps its most famous role: that of the development of radar. However, radar only answered the need of early detection; word still needed to get out to the defensive batteries and pursuit plane airfields, and that word would travel by wire. At work in all of the military planning of the interwar period was the isolationist sentiment strongly entrenched in the country. Any future war, it was said, would surely be a war for the defense of the American homeland. The nation was through with foreign wars, especially those started by the Europeans. Thus, plans for aircraft warning and defense only dealt with the coastal regions of North and Central America; a very extended region, but one that held the benefit of well-established communications systems. In any future war, it was believed by war planners, the military would be able to take advantage of the nationwide system of wire-based communications.[10] In fact, in terms of an aircraft warning system, the public telephone and telegraph utilities were expected to play a prominent role. Throughout the 1930s, exercises were held in which these utilities and their employees took part. Though the response times for aircraft warnings were not as good as had been hoped for (up to five minutes from detection to alert in some cases), the utilities saw them as a success and called for government assistance in increasing the sizes of their companies in the name of national defense.[11]

Though of dubious success, these exercises only demonstrated the effectiveness of the communications systems in those areas (the East Coast and coastal Southern California) where the population density called for extensive networks. Maneuvers carried out by ground forces during the 1930s in such underpopulated parts of the country as Texas and Louisiana pointed out the potential problems of operating in areas without extensive commercial resources. Early in the decade, these maneuvers only seemed to prove the point being made by the Signal Corps: research and development work needed to be concentrated on ways to improve the means of a ground force for setting up its own wire-based system when operating in territory lacking sufficient resources. However, as the decade wore on, and the armored forces began to steer military thinking toward the twin goals of mechanization and mobility, the roles of radio and wire, in the minds of the ground forces, began to reverse. By 1940, things had pretty much come to a head due to two primary factors: the failures of wire technology and the successes of radio in the 1940 summer maneuvers in Louisiana, and the advent of FM radio technology.

By 1940, the picture of what the next war would be like had become much clearer. The German air force, utilizing the Spanish Civil War as a training exercise, had demonstrated the close cooperation possible between air forces and mobile ground forces. Hitler's rapid conquest of continental Europe during 1939 and 1940 laid to rest any remaining doubts as to the importance of mobile forces interconnected through radio. In May 1940, 60,000 U.S. troops were assembled along the Sabine River of Louisiana for maneuvers that, more than any other, would focus on the mobility needed in modern warfare and also, more than any other, would point out the serious shortcomings of wire-based communications.[12] Even though the area did possess commercial resources, communication between units was hampered. Field wire failures were many; some due to the amount of vehicle traffic rolling over the wires, some due to rainy conditions, and some due to the preference shown by farm animals for their apparently tasty insulation. It appears that some types of insulation were preferred by cows and others by pigs; a Signal Corps lieutenant jokingly suggested that the type of wire used in a particular area be dictated by the type of farm animals present.[13]

Still, there were those in the Signal Corps who held to the wire-over-radio status quo, if for no other reason than in the interest of military secrecy. Wire-based communication, many Signal Corps officers believed, was secret (see Chapter 3 for counterexamples of this). Radio, however, was out in the open for anyone to receive. Radios in the hands of the common soldier, especially those capable of sending voice transmission, were quite possibly "more dangerous than useful."[14] Be that as it may, in the minds of those whose job it would be to use the radios in combat, it was high time for a change. And, ultimately, the Signal Corps was forced to make this change, in large part due to its own developments in the areas of frequency control and FM radio.

Perhaps the most important detail in radio technology, whether for transmitters or receivers, is frequency control, i.e., how the radio is constrained to transmit on, or receive, the particular frequency of interest. Distilled to its most basic concepts, a radio works because the electrical current within a particular part of its circuitry is oscillating. The frequency at which it is oscillating determines the transmitting or receiving frequency. All things in nature that are able to oscillate (i.e., vibrate) possess what is known as a *natural* or *resonant* frequency. Left to its own devices, this is the frequency at which an oscillating system will vibrate; a violin string, for instance, will vibrate at its natural frequency when plucked and will therefore give off its characteristic musical tone. Just what the natural frequency of the system is depends on its physical characteristics. For an object oscillating at the end of a spring, it depends on the stiffness of the spring and the mass of the object. For a simple pendulum (or, approximately, a child on a playground swing), it depends on the length of the pendulum. For a violin string (or that of any other stringed instrument), it depends on the thickness and the length of the string.

No matter what the physical system, if it can oscillate, something about it determines its natural frequency. Electronic circuits are no exception. Perhaps the simplest type of oscillating electronic circuit can be constructed from a capacitor, an inductor, and a resistor (available for pennies at any electronics store). A capacitor is an energy storage device. By "condensing" positive and negative charges on opposing metal plates separated by an insulator, energy is stored in the electric field existing between the plates. Capacitors only allow current to flow when they are either charging (collecting charges) or discharging (releasing charges). Thus, a completely charged or a completely discharged capacitor will stop the flow of current. Under alternating currents of high frequency, a capacitor is never allowed to charge or discharge very much, thus offering little "resistance" to the current flow. An inductor is little more than a coil of conducting wire. However, it has the ability to store energy within a magnetic field generated by rapidly alternating currents. In opposition to a capacitor, it only allows alternating current to flow easily when the frequency of the current is low. A resistor is a component that dissipates energy from the circuit as current passes through it, regardless of which direction or at what frequency that current is traveling. How easily current flows through a circuit containing all three of these components depends upon their mutual interaction. It turns out that current flows best through such a combination of components at one particular frequency: the resonant frequency. Thus, in this respect, alternating electric circuits behave exactly like all other physical oscillators.

Used within a radio circuit, this simple combination of electronic devices determines the natural oscillating frequency of the radio, and therefore controls its transmitting and receiving frequency. Should the physical conditions of an oscillating system change, its frequency of oscillation will change (a fact learned by most children around the age of three years old when they learn to swing by themselves by "pumping" their legs). The electrical characteristics of the capacitors, inductors, and resistors in a simple radio circuit depend on such things as temperature and humidity. Thus, a radio based on this type of simple circuit will change its frequency as it heats up or cools down, or as the weather conditions change. Throughout the first third of the 20th century, efforts at *stabilizing* the frequencies of radio were carried out. By the late 1930s, two primary frequency control options existed: the master oscillator concept (based on the previously described combinations of electrical components) and a radically different technique known as quartz crystal control.

A great deal of progress was made in oscillator technology in the years approaching the Second World War. Primary among the goals of radio engineers was a method of eliminating, or at least compensating for, the drift in frequency due to changes in temperature. Special circuits were developed to compensate for this "thermal drift" in the tuning circuits. In other cases, the tuning components were manufactured from materials found to be relatively unaffected by temperature changes. One other option was to enclose the entire oscillating portion of the circuit within a constant-temperature oven. Furthermore, the oscillating portions could be isolated or buffered from the effects of the surrounding circuit by use of amplifiers. These "master oscillator power amplifier," or MOPA, sets were admirably stable, provided they were given a nominal 20 minutes to warm up and come to thermal equilibrium. They also offered the benefit of an almost unlimited number of frequencies at which they could function; tuning being accomplished by adjusting the variable components within the tuning circuit. They were far from perfect, however, as will be discussed in detail in the following chapter (for instance, they could be nearly impossible to tune while in a moving tank, jeep, or armored vehicle).

The other primary option for frequency control during the pre-war years was the use of quartz crystal oscillators, also known as quartz crystal units (QCUs). The QCUs were composed of thin wafers of quartz crystal placed between metal plates (which served as electrodes). By virtue of the piezoelectric (pronounced *pie—ezo—electric*) effect, quartz behaves as a natural oscillator when placed within an electric circuit, with the natural frequency of the quartz wafer governing the oscillating frequency of the circuit. The generation of electrical "polarization," or voltage, through mechanical deformation is known as the direct piezoelectric effect and was discovered by Jacques and Pierre Curie in 1880. A household application of this effect can be found in some types of lighters used to start fires in grills or fireplaces. When the trigger of the lighter is pulled, pressure is placed on a small piezoelectric crystal, producing the needed voltage (sans batteries) to power the starter. Piezoelectric crystals, particularly quartz, are also used in pressure-sensing devices. Changes in pressure lead to changes in the induced voltage across the crystal, which is monitored electronically. Such devices prove to be extremely sensitive and physically robust instruments. The converse piezoelectric effect, predicted by G. Lippmann in 1881 and later observed by the Curies,[15] is physical deformation due to the application of a voltage across the crystal (such as by placing it in an electric field or within an electronic circuit). This physical deformation can take the form of a bending, shearing, or torsion (twisting).

Piezoelectric crystals are quite common; of the 32 known classes of crystals found in nature, 20 of them exhibit piezoelectric properties.[16] Of these 20 classes, quartz is by far the most suited for electrical oscillator use, having the best combination of piezoelectric along with electrical, mechanical, and thermal properties. Extreme hardness, for example, is just one physical property required of piezoelectric oscillators; oscillating at 100 MHz (i.e., 100 million oscillations per second), a point on the surface of a crystal oscillator is undergoing accelerations several million times that of the acceleration of gravity (several million "g's" in fighter plane parlance).[17] Substances of lesser internal strength might very well fly apart under such conditions.

The piezoelectric effect results primarily from the disturbing of the precisely ordered planes of atoms within a crystal. Electrically neutral when in an undisturbed state, the displacement of the atomic planes leads to polarization of electrical charge within the crystal (i.e., a separation of the more positively charged atoms from the more negatively charged ones) and the production of an electrical "potential difference" or voltage between opposite faces of the crystal. In the converse effect, the atoms within the crystal physically move to realign themselves with an externally imposed electric field. The particular movement of the atoms depends on the characteristics of the electrical field. Reversing the polarization of the field (i.e., making the positive direction negative and the negative direction positive) results in a similar reversing of the internal structure of the crystal. An electric field that continuously switches its positive and negative orientation will result in the crystal constantly realigning its internal structure (a physical oscillation of the crystal).

As discussed above, all physical oscillators display preferences for particular manners and frequencies of vibration. Oscillators vibrate most efficiently and with the greatest amplitude when in these "resonant modes." Consider a quartz wafer placed between two electrodes and incorporated into an electric circuit (such a configuration would be referred to as a quartz resonator). The presence of an alternating voltage will cause physical oscillations to occur within the wafer (through the converse piezoelectric effect). These physical oscillations, however, will lead to the generation of an alternating voltage across the wafer (through the direct piezoelectric effect). Thus, the wafer can act as a conduit for the voltage signal propagating through the circuit. However, the efficiency of this transmission depends greatly on the frequency of the imposed voltage. If the frequency of the alternating voltage is far removed from the natural or "resonant" frequency of the crystal, the oscillations of the wafer are extremely weak and the quartz acts much more like an insulator, impeding the transmission of the signal. When the frequency of the signal is at or very close to the resonant frequency of the crystal, the wafer's oscillations are at a maximum and the signal is transmitted very effectively. Thus, the crystal serves to control the frequency of the oscillator, only permitting it to oscillate efficiently at the resonant frequency of the crystal.

Academic research involving quartz crystals as frequency control devices had been carried out for some years before World War II. As a student at Wesleyan University during the 1920s, Karl Van Dyke was involved in research on crystal oscillators. Due to the manner in which quartz oscillators reacted to alternating voltages and currents and how they stored and dissipated energy, Van Dyke concluded that their electronic characteristics could be described in terms of an "equivalent circuit" composed of two capacitors, an inductor, and a resistor. This equivalent circuit (the diagram of which is shown in Figure 1.1) allowed electrical engineers to easily determine the effects of placing various oscillators into radio circuits and enabled them to design oscillators to have just the types of characteristics needed.

The resonant frequency of a quartz wafer depends primarily on its physical dimensions. In the case of the shear mode of vibration, the most important dimension is the thickness. The thicker the wafer, the more mass and inertia the wafer possesses. Thus, a thicker wafer will vibrate more slowly, or with a lower frequency, than a thinner wafer. To produce a crystal oscillator of a particular frequency, the wafer must be ground to a very precise thickness. In order to create a resonator, slabs of quartz are initially cut from a "mother crystal" and then sliced into relatively thick wafers (millimeters thick), usually with diamond-tipped saws. "Blanks" are then diced from these irregularly edged wafers using a trim saw and then squared using a diamond-impregnated grinding wheel.[18] Blanks are then "rough ground" (or "lapped") to near the final thickness (often within 0.4% of the desired frequency).[19] Ultimately, it was determined that a three-stage lapping procedure, each stage utilizing a finer grade of abrasive than the one before, worked best.[20]

The final lapping steps needed to bring the blank to the desired frequency are referred to as "final finishing." Though the rough grinding can be done by machine, the final lapping is done by hand, moving the blank around a grit-coated glass plate with a fingertip.[21] As the frequency response of the crystal blank is extremely sensitive to its mass, only a tiny amount of surface quartz may be needed to be removed during the final finishing step to bring the blank to the proper frequency. If too much quartz is removed, the frequency of the blank will be too high (and may now be worthless to the manufacturer). Once a blank is finished to final frequency, great care has to be taken to maintain the cleanliness of the blank. Any foreign material (including oil from the finisher's skin) would add mass back to the blank and lower the oscillating frequency. The production of crystal oscillators was an activity requiring a high degree of precision, accuracy, and attention to detail.

Once ground to final frequency, the crystal blank would be cleaned, dried, tested for activity (a measure of its ability to vibrate), have its frequency double-checked, and then be placed within the "holder." The holder was designed to both protect the crystal blank and allow it to be incorporated into

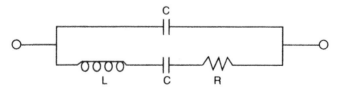

Figure 1.1. The equivalent circuit of the quartz oscillator

the radio for which it was designed. Most World War II–era holders were molded of material such as Bakelite and included two plugs or pins that allowed them to be connected to the electric circuit within the radio set. Though quartz crystal units came in a wide variety of sizes and shapes, two of the most common ones will be described here. The first, the FT-243 used in most vehicular radio sets, was small; approximately 2 cm wide, 3 cm long, and 1 cm thick (see Figure 1.2). The metal face plate of the unit carried the essential information of the oscillator: the unit type, the frequency in kHz (kilocycles during the World War II era), the manufacturer, and the unit's serial number. Most face plates also contained a space for the noting of a channel number, but, as this applied to the particular operating scheme of the military unit utilizing the QCU, this information was not always included.

The face plate of the FT-243 was held on by three small screws passing completely through the unit from front to back. Under the face plate, a small gasket served to protect the interior region from water and other foreign

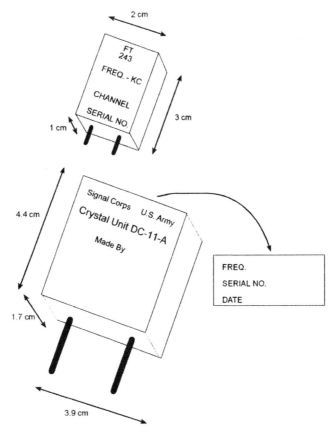

Figure 1.2. Crystal holders FT243 (*top*) and DC-11-A (*bottom*)

material. Beneath the gasket was a spring-like mechanism that also served as an electrode (connected to one of the holder's pins with copper wire). A similar mechanism, connected to the other pin, was soldered to the back of the holder. Between the two spring electrodes was the quartz blank or "plate," sandwiched between two metal plates. One of the metal plates was machined so that it only contacted the plate at its center and corners (thus leaving the remainder of the plate free to vibrate). The quartz plate itself was a rectangle 1.5 cm by 1.2 cm and 0.33 mm thick.

The other commonly used crystal unit during World War II, the DC-11-A used in most crystal-controlled aircraft radios, was of similar design. It was larger than the FT-243; 4.4 cm long, 3.9 cm wide, and 1.7 cm thick (see Figure 1.2). The unit also differed in that the unit type and manufacturer information were printed on the front, but the frequency and serial number of the unit were stamped into a metal plate at the top. Below the metal plate (held on by two screws) were two more screws that allowed the top of the unit to be opened. A spring-like piece of folded metal held the metal-quartz sandwich tight against the front of the holder. The front and back of the holders consisted of smooth metal plates serving as electrodes, connected to the holder's pins. The quartz plate of the DC-11-A was square, 1.8 cm on a side and 0.31 mm thick. (See plates for images of assembled and disassembled units, including the FT-243 and the DC-11-A.) Deceptively simple in their design and construction, nevertheless, quartz crystal units presented some of the most difficult manufacturing challenges of any wartime commodity.

The ability of a system to maintain its oscillations is oftentimes a very important characteristic. Oscillating systems are often described in terms of their energy losses utilizing a term known as the "Q factor" (defined as the total energy stored in the oscillator divided by the energy lost during each oscillation cycle). An efficiently oscillating system (one possessing a high Q value) is one that suffers little energy loss. On the other hand, the oscillations of an inefficient, or low Q, system damp out fairly quickly. It is not unusual for quartz crystal oscillators to have Q values ranging from hundreds of thousands to millions (compared with, say, a piano string, which might possess a Q value of only a few thousand).[22] In practice, this means that the frequency of a radio utilizing a crystal oscillator for frequency control suffers almost no drift in its operating frequency. This characteristic alone made quartz crystal units immediately popular with amateur radio enthusiasts when they arrived on the commercial scene during the mid-1920s.

Quartz crystal units had serious drawbacks, however. First, the raw quartz from which radio oscillators could be produced was rare; the primary source of radio-grade quartz was the remote Minas Gerais region of interior Brazil. Whether sufficient quantities of raw quartz could be obtained in time of war to supply the military with the oscillators it would need was the single-most problematic issue facing the Signal Corps in its pre-war frequency control deliberations (it is also the topic of Chapters 6–8). Secondly, though crystal control offered unparalleled frequency stability, it restricted a radio to a single frequency; to change to a different frequency, a different crystal unit had to be plugged into the radio.

For the Signal Corps in the last years before WWII, the question was this: whether to accept the master oscillator method of frequency control or to gamble everything on the greater stability and ease of operation provided by quartz. One offered a "known" and easily manufactured technology but did not allow for the mobility the ground forces desperately needed. The other gave impressive stability and "push-button" tuning yet might be impossible to mass produce in time of war. The decision of which technology to use would have great implications for the entire war effort.

2

CRYSTAL CONTROL—THE GREAT GAMBLE

The use of quartz for radio frequency control had a long and established history. Though other types of piezoelectric substances had been shown to be useful for controlling the frequencies of electronic oscillators, Walter G. Cady, of Wesleyan University, first demonstrated in 1923 that quartz resonators could be used very effectively to control the transmitting and receiving frequencies of radio sets.[1] From this discovery until its grudging acceptance by the Signal Corps in late 1940, quartz crystal control existed primarily in the world of the amateur radio enthusiast. It was the "ham" radio operator who first seized upon crystal units as a useful means of frequency control. Early hams, by their very nature, were experimenters, many finding no more joy than in building and constantly improving their homemade "rigs." Many amateurs carried this enthusiasm for the homemade into the field of crystal control—designing, building, and, in some cases, ultimately manufacturing and selling their crystal units to other hams.[2]

Most early hams adopted crystal control as a means of stabilizing their transmitters. In the earliest transmitter designs, the antenna was a part of the tuning circuit. If, for instance, the antenna moved in the wind, the frequency of the transmitter would change. These radio sets were also very susceptible to temperature changes. In order to maintain their assigned positions in the increasingly crowded radio spectrum of the early 1920s, hams bought or made crystal units.[3] The ham radio operator was helped in this endeavor by the radio publications of the day. The primary publication, *QST*, carried advertisements for relatively cheap ($4) unfinished quartz blanks (which hams could grind by hand to their own particular frequency) as well as for rather expensive ($35–$50) oscillators ground to specified frequencies.[4] Along with advertisements came "do-it-yourself" articles, circuit schematics, and glowing testimonials for the new technology:

> Can you imagine a transmitter that never shifts its wavelength even a hundredth of a meter? Can you imagine making a schedule for 96.38 meters and knowing that you will be right on that wavelength and knowing that the other man will be tuned right to you? These things are possible with the oscillating crystal.[5]

By 1926, commercial radio stations were making the switch to crystal control. Station WEAF in New York City was the first, but, within a few years, all commercial radio stations in the United States were crystal controlled.[6] For amateurs and commercial outfits alike, being able to depend on the stability of their transmitter frequency was essential. For amateurs, their enjoyment of their hobby was at stake; for the commercial stations, their financial success. In terms of the market for crystal oscillators, the commercial stations had very little effect or influence. Since commercial broadcasters only transmitted on a single frequency, they rarely needed more than one operating unit and one spare. Hams, on the other hand, needed a larger collection of crystal units to give them flexibility in receiving frequency. This, coupled with the overwhelming numbers of amateurs compared with commercial stations (see Figure 2.1), meant that the crystal producers catered to the needs of the amateurs almost exclusively until the latter 1930s, when the "two-way" radio began to be marketed to police and fire departments, taxi companies, and

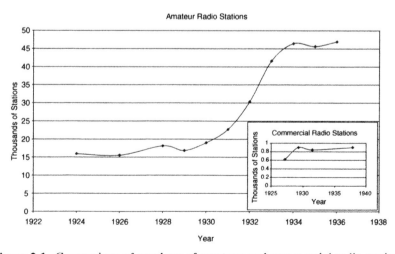

Figure 2.1. Comparison of numbers of amateur and commercial radio stations

commercial airlines.[7] The Paul Galvin Corp. (later renamed Motorola) entered the radio communications business in 1936 with its *Police Cruiser* AM radio utilizing crystal control for its single operating frequency. By the following year, all of its car radios featured push-button tuning. In 1940, an improved AM two-way set was being used by police stations around the country, and, by 1941, Galvin was marketing a much superior FM set. Crystal-controlled mobile radio had arrived on the civilian scene.[8]

"IT HAS BEEN THE POSITION OF THIS LABORATORY THAT CRYSTAL CONTROL WAS A SOLUTION TO BE EMPLOYED ONLY AS A LAST RESORT"

Though the military's association with quartz oscillators predated even that of the ham radio community, it took much longer for it to accept crystal control at the level of the amateur and commercial communities. One of the earliest uses for the piezoelectric properties of quartz was in an early version of SONAR. During World War I, Professor Paul Langevin of France, in an effort to combat the German submarine menace, used quartz crystals in an instrument designed to produce and detect sound waves under water.[9] Similar research was instigated in the United States by the War Department, but the end of the war brought a quick end to the work.[10]

The utilization of quartz crystals by the Army Signal Corps progressed slowly throughout the 1920s. A meter for measuring frequency was developed in 1921 that maintained calibration through the use of a crystal. After Cady's 1923 success with radio frequency control, the Army and Navy's signal laboratories began to experiment with crystal controlled radios. During 1927 and 1928, a number of transmitting stations utilizing crystal control were constructed throughout the country as part of the War Department Radio Net.[11] At this time, the crystal oscillators were somewhat susceptible to thermal drift, so their use was confined to fixed-site facilities that could be maintained at constant temperature. However, after the discovery of the technique by Bell Labs scientists for producing crystals whose oscillation characteristics showed very little or no dependence on temperature in 1934, this restriction was removed and experiments with crystal control in mobile radio sets began in earnest.

As discussed in Chapter 1, the road toward military acceptance of crystal control was a long and contentious one. One reason for this is that the Signal Corps could never reach an official decision as to its own feelings on the issue. The Chief Signal Officer (CSigO) received contradictory recommendations from various Signal Corps laboratories and from various branches of his own office. In 1935, the CSigO received memos from both the Aircraft Radio Laboratory (ARL) and the main Signal Corps Laboratory (SCL) at Ft. Monmouth, New Jersey. The ARL memo, though admitting that crystals gave much greater stability than any other means of frequency control, stressed that this method should only be used "as a last resort." They felt that once the decision was made to utilize crystal control, the Signal Corps would have started down a very slippery slope. If transmitters were outfitted with crystals, then why not receivers? And, in order to have tuning flexibility, a large number of crystals would be needed for each and every one of these transmitters and receivers. How then, asked the ARL, could the Signal Corps hope to support such a program when the current production numbers for QCUs in the United States were "pitifully inadequate?"[12] Similar questions were still being asked in the spring of 1941.[13] The other side of the coin was laid out in a memo from the Signal Corps Lab that July. Here the argument was that the overwhelming increase in stability and *ease of use* far outweighed any doubts about supply. In fact, the memo went so far as to suggest "the belief that the problem of war supply is far from impossible of solution."[14]

The arguments against crystal control were built around three primary points: lack of flexibility in tuning, worries about inadequate supplies of raw quartz, and the attendant cost of outfitting radio sets with the needed crystals. In the spring of 1940, the Air Corps requested that the SCL carry out a test using crystal control in the SCR-183 command radio set. The response to the Air Corps from the Office of the Chief Signal Officer made use of all three of the above arguments. Colonel Clyde Eastman, writing for the CSigO, first pointed out that the current master oscillator-controlled SCR-183 functioned just fine after "the usual warm-up period of 20 minutes" and that, after this warm-up period, the sets should experience "but little variation" in frequency. He then went on to outline an eight-part argument for why their request should be denied. First, no test should be carried out because the conclusion was already known: the crystals would provide much greater stability than the current circuitry. This would then lead to a request on the part of the Air Corps for crystals for all of its command sets. At 20 crystals per set (to provide adequate tuning flexibility), and an estimated $10 (peacetime) or $15 (wartime) cost per crystal, it would cost an additional $40,000–$60,000 to produce the 2,000 SCR-183 sets believed to be needed the following year.[15]

Colonel Eastman then conveyed the "alarm" with which the CSigO viewed any proposals for increased crystal control. It was feared, he said, that the **present demand** (boldface in original memo) for crystals may not be met by the manufacturers, much less an increased demand. If there were problems with supply and distribution of crystals under the current conditions, what could they expect with increased requests? The Navy, he pointed out, appeared to be moving away from crystals and back to the master oscillator technology whenever they could. This was due in part to the progress in master oscillator technology. Yes, the master oscillator sets required greater care in tuning, Eastman admitted, but with proper training of the operators, that should not be a problem. In his final two points, Eastman used the Air Corp's own reports against them. In a 1938 report, the Air Corps Technical

Committee concluded that, "Crystal controlled radio equipment is too inflexible to meet war-time demands and even if it could be made flexible enough by procurement of sufficient crystals it would be exceedingly costly for tactical purposes." Furthermore, the Chief of the Air Corps himself had recently called for a freeze in the design of radio equipment. This current request for a test of crystal control in the SCR-183 would clearly violate this policy.[16]

Flexibility, supply, and cost were three very strong arguments against crystal control (provided the new master oscillator-controlled sets were as good as advertised). Colonel Eastman may have been singing higher praises for the master oscillator sets than they deserved. A January 1940 report from the SCL, entitled "Comparison of Master Oscillator and Crystal Control for Field Radio Sets," gives a much different impression. Master oscillator-controlled sets did appear to work well, provided they were given adequate time to warm up and were not in a moving vehicle. However, the report pointed out that, under "adverse conditions," these sets might "preclude satisfactory net operation" (i.e., a collection of radios all operating on the same frequency). Frequency meters could be used to monitor the master oscillator–controlled sets and maintain their calibration. However, such checks might need to be carried out several times a day.

While only requiring five minutes for a skilled operator, provided the radio set was not in motion, such maintenance might well be impossible in a moving vehicle. As crystals were found to be "10 to 20 times" more accurate than a master oscillator set, they also allowed for 5 to 10 times as much communication (in terms of equal operating times). These benefits, however, could only be guaranteed if the set were equipped with its full complement of crystals. The report ended on a note reminiscent of the SCL 1935 memo, pointing out that, even though the only source for radio-grade quartz was Brazil, it was a country linked to our own by "long-standing friendship, the Monroe Doctrine, the 'good-neighbor' policy, our hemisphere defense policy, and fast and cheap air and water communications."[17]

This pattern of requests for crystal control from the combat branches being supported by SCL and opposed by the Office of the CSigO (OCSigO) continued throughout 1940. The primary instigator of these arguments during this time was the armored forces. In late 1939, the Infantry had requested field tests of crystal-controlled radios for its light tank units. Upon hearing this, the Mechanized Cavalry (the other half of the as-yet unconsolidated armored forces) asked to run its own tests as well.[18] In both cases, a sizable number of crystals would be needed, along with the assigning of frequencies for their use. All of this had to be approved by the OCSigO. The Infantry made a very good case for the extra crystals: their current radio situation consisted of sets containing four crystals along with a master oscillator circuit. Whereas four crystals (and therefore four communication channels) were adequate for small units, it was far too few for large unit operations. The request for the crystal field tests was approved in December 1939.[19] The approval process for the Mechanized Cavalry's March 1940 request was a little more contentious, however. A representative of the Communications Liaison Division within the OCSigO stated, "I am opposed to the entire project for either arm [Infantry and Mechanized Cavalry], for unless we are prepared to stock crystals for all requirements up to M [mobilization day] plus 1 year, we are letting them use and become accustomed to something impracticable in wartime."[20] This argument would be repeated in another Signal Corps report 12 months later.[21] The request of the Mechanized Cavalry was eventually approved in April 1940.

As with Colonel Eastman's predictions for the proposed Air Corps tests, the armored units were pleased with the crystal tests and wanted full crystal control for all of their units. This led to further consternation on the part of the OCSigO. In mid-August of that year, in a memo to CSigO Joseph Mauborgne, Colonel Hugh Mitchell stated that it was the opinion of his Research and Development (R&D) Division that "the flexibility requirement in tactical radio sets renders the procurement of a sufficient number of crystals impractical at this time." With the armored forces asking for upwards of 400 frequency channels, their crystal needs simply could not be met. To his further irritation, *his pleas for a reduction in the requested number* of channels were met with requests for even more! Mitchell made the point that the best compromise was to combine a small number of crystals with master oscillator circuits. He felt that the increase in stability gained through total dependence on crystals was not worth the cost. Plus, according to his information, most of the frequency drift problems with master oscillator sets were due to operator error, not the equipment (this was the forerunner of today's arguments between software users and software designers). Mitchell further took the SCL to task for encouraging, or at least favoring, increased crystal use. Referring to Colonel Eastman's memo, Mitchell felt that the attitude of the SCL was "inconsistent with the spirit" of the CSigO's alarm at further requests for crystal control.[22]

One week later, the War Plans and Training Division of the OCSigO stated its agreement with the Mitchell position. Their argument centered on the operational drawbacks of crystal control—primarily insufficient tuning flexibility. What would happen, they asked, if a communications channel were "jammed" by an enemy? With a master oscillator set, the operator would simply retune to a nearby frequency and continue to transmit. With a crystal-controlled set, however, this would be impossible. Whereas broadband interference would be needed to completely jam master oscillator–controlled communications (likewise interfering with the enemy's own transmissions), an enemy could selectively jam crystal-controlled frequencies while leaving its own channels open. Furthermore, they pointed out, the crystal units made the radio equipment more complicated and prone to failure. In their opinion, the master oscillator sets were now stable enough to provide for operation of transmitters on neighboring channels much closer in frequency than ever before.[23]

"IF WE ARE TRYING TO GET PRACTICALLY AUTOMATIC RADIO SETS AT THE PRESENT TIME, ONLY CRYSTALS FILL THE BILL"

Throughout all of the deliberations over crystal control up to and even after Pearl Harbor, the OCSigO seemed to treat crystals as a forbidden fruit. Though their great promise enticed and tempted the using arms and the development labs alike, the OCSigO felt that the promise of crystals would be difficult if not impossible to keep once the United States became involved in the war. The R&D Division within the OCSigO was in a particularly troubling position: on the one hand, its development laboratories were very much in favor of crystal control; on the other, it had the responsibility to consider the supply side of the equation. In a subtle bit of irony, the 1940 head of the SCL, Colonel Roger Colton, ultimately came to head all research and development activities as a Major General in command of the Engineering and Technical Service under CSigO H.C. Ingles. The decisions he made and the policies he supported as head of SCL in 1940 would come back to haunt him later in the war when the supply situation became very critical.[24] In fact, the effort to create a mass production industry for QCUs and to supply it with raw quartz would come to dominate his life throughout the war. However, in 1940, these details were still part of the unknowable future; Colton's job, as he saw it, was to help the using arms anyway he could by developing the stable, dependable, and mobile communications equipment they needed.

Colton was a good choice for the position. Educated in electrical engineering at Yale and the Massachusetts Institute of Technology (M.I.T.), he had always been associated with the research and development activities of the Signal Corps. Along with having published scientific articles on the physics of radio, Colton was a long-time believer in the military benefits of wireless communication. His primary goal during the last years of the 1930s was to make believers of both the combat branches and the Signal Corps leadership.

To demonstrate the best of the new radio technology, Colton visited major ground force headquarters across the country. In the fall of 1939, his road show was at Fort Knox, Kentucky, headquarters of the Mechanized Cavalry Board. Though his equipment was still somewhat primitive, he was able to demonstrate the SCR-245, the first truly mobile, crystal-controlled radio useful for armored vehicles and jeeps.[25] The SCR-245 had been modified to utilize crystal control in addition to its original master oscillator circuitry at the request of the armored forces (the master oscillator's seven different tuning regulators made it nearly impossible to use on the move).[26] In addition to the SCR-245 and a truck-sized transmitting station, Colton was also able to demonstrate a very early model of an FM set being developed for aircraft use.[27]

The stories of crystal control and FM are nearly inextricable. In fact, they represent something of a chicken-and-egg situation: without crystal control, tactical FM radio would not really have been feasible, but without the promise of FM technology, the support of the ground forces that led to full-scale development of crystal control would not have existed. The development of FM radio technology by Dr. Edwin Armstrong represented a tremendous step forward for radio and the answer to the prayers of the mobile ground forces. AM, or amplitude-modulation, radio suffers a great deal of interference due to such things as atmospheric conditions (humidity, thunderstorms, etc.), "cross-talk" from stronger transmitters on neighboring frequencies, and the constant pop and static due to radio emissions from automobile spark plugs. The last source, more than any other, made AM radio nearly impossible for use in moving vehicles.[28] (Note: the air forces, much preferring the longer-range operation of AM technology, devoted great effort to shielding against this engine-induced interference.[29]) FM radio utilizes frequency modulation, changes in frequency, to carry information (as opposed to the changes in amplitude or signal strength utilized by AM radio) and is thus immune to the noise and static that plague AM receivers. Furthermore, FM radio operated at much higher frequencies, above the AM bands then becoming crowded due to increased radio traffic.

Another very useful benefit of FM was the fact that it allowed multiple transmitters to utilize the same frequencies without interference. Whereas an AM signal might need to be 20 times as strong in order to overcome another interfering signal on the same frequency, an FM receiver would lock onto a signal only three times the strength of the interfering signal.[30] For large units operating in close conditions, this represented a great advantage in that the number of needed frequencies (and likewise the needed crystals within the sets) could be reduced. This also meant that FM sets were much less suscep-tible to jamming. After the war, a veteran of combat in Europe said of FM: "I know the fighting would have lasted longer if we hadn't had FM on our side. We were able to shoot fast and effectively because we could get informa-tion quickly and accurately by voice, on FM. FM saved lives and won battles because it speeded our communications and enabled us to move more quickly than the Germans, who had to depend on AM."[31]

From the armored forces point of view, crystal-controlled FM sets were everything they had hoped for. Particularly important was their immunity from interference from vehicle engines. As a combat arm, the Armored Forces were envisioned to take advantage of "tactical swiftness." Their com-munications had to be just as mobile ("one hundred miles in motion" was the catch-phrase coined by Colonel Dawson Olmstead (future CSigO)).[32] In combat, they could no longer wait 20 minutes for their radios to warm up, nor bring their vehicles to a halt to use them, than they could depend on telephone wires running to their tanks and jeeps. In fact, when questioned by CSigO Maubornge as to the wire needs of a mechanized division, Brigadier General A.R. Chafee (commander of the Armored Forces) replied that, if used at all, wire would be restricted to nighttime halts and rear-echelon rest areas.[33] Thus, General Chafee and Colonel Colton joined forces, determined to con-vince the OCSigO to accept large-scale crystal control.

In fact, to his surprise, Colton discovered at his 1939 Ft. Knox exhibition that officers within the Armored Forces already knew about crystal control and took almost no convincing on his part to accept it. Radio enthusiasts within the Mechanized Cavalry had discovered the use of crystals in a small two-way radio set manufactured by the Link Corporation and in use by a few police departments around the country.[34] Colton had to point this out to the OCSigO in an August 1940 memo after being accused of too-strongly advocating crystal use to the Mechanized Cavalry. Once on the table, however, he did not shy away from endorsing the Cavalry's request. The real problem for the ground forces, in Colton's view, was not too many crystals, but too few. Supplying field radios without the full complement of crystals, as was currently being done, was about as bad as not providing crystal control at all. Adequate numbers of crystals needed to be supplied ("A crystal set can't be divorced from its crystals," Colton wrote).

He also pointed out the fallacy behind the claim that master oscillator sets were just as stable as crystals *after being allowed to warm up*. Whereas the air units had ground crews to warm up the engines and tune the radios on their airplanes before a flight, the armored units had no such luxury. Furthermore, the claim of added cost due to crystals was overblown.[35] In referring to a report concerning "Relative Costs of Master Oscillator and Crystal Control Radio Sets, Type 3, for Armored Forces," he pointed out that, with the advent of smaller crystals, the costs of crystals could be brought down to the point ($5 apiece) where they represented no more expense than master oscillator sets. That, added to the benefits of crystals with regards to time of development (two to five times less), protection against interference and noise (three to four times greater), extent to which current designs could be utilized (75%–100%), and the ability to meet or exceed current military demands argued very strongly in favor of utilizing crystals.[36] "If we are trying to get practically automatic radio sets at the present time," Colton said, "only crystals fill the bill."[37]

As early as January 1940, Colton had been pressing the OCSigO on another of the benefits of crystal control: accurate and stable assignment of channels within a frequency band. The range of frequencies assigned to field units were not very wide (as small as 260 kHz in the case of the armored units).[38] In order to operate radio "nets," the assigned frequency range had to be divided into channels. The minimum allowable frequency width between adjacent channels and the ability of a radio to remain within a designated channel determined the number of channels available to the unit. By summer of that year, with the formation of the Armored Forces combat arm, General Chafee was adding his more powerful voice to the call for greater numbers of channels, and for FM radios to utilize those channels. Research and Development, attempting to hold the line against increased crystal control, was against this proliferation of channels; a situation, in the words of Colonel Mitchell, surely to result in "channel congestion and interference."[39] Mitchell's arguments carried little weight, however. Just as the armored forces had helped drag tactical radio into full use, it would also do the same for crystal control. A dramatic policy shift was about to take place within the OCSigO, and men such as Colton and his SCL engineers were going to have to shoulder the burden of supporting it.

Before such a policy shift could take place, the long-held fears of the OCSigO about the availability of raw quartz for manufacturing QCUs had to be addressed. Colton and his executive officer, Major James D. O'Connell, in an attempt to settle once and for all the questions regarding supply, held a conference on August 8, 1940, with representatives of Western Electric, the manufacturing arm of Bell Laboratories and the country's largest crystal oscillator producer. One of the representatives, C.R. Avery, informed the Signal Corps men of the quartz mining conditions in Brazil and how, currently, the U.S. crystal manufacturers were only able to utilize a small percentage of the raw quartz mined. This was due to the fact that a great deal more labor (and thus cost) was involved in working with the lower-quality crystals that made up the majority of the yields from the Brazilian mines.

In terms of quantity, Avery felt that, using current methods of purchase and import, up to 30,000 pounds of raw quartz per year should be available. With government participation, the amount might reach 40,000 pounds; higher still if Great Britain could be persuaded to "drop out of the market." Avery pointed out that if the techniques developed at Western Electric for working with the more difficult unfaced "river quartz" were adopted industry-wide, up to 60,000 pounds of raw quartz might be available each year. Discussion next turned to the current stockpiles of crystal manufacturers, amateur radio operators, importers, and even museums. A conservative estimate from this survey was that a little over a half million crystals could be obtained from these sources. The conference ended with a production estimate of between 1.5 and 2.4 million crystals per year with raw quartz imported from Brazil.[40]

In September 1940, the decision was made: development would begin on a series of crystal-controlled vehicular radios (named AF-I, AF-II, AF-III, and AF-IV) for use by the Armored Forces. These would comprise both long-range (100 miles) and short-range (less than 5 miles) sets. The short-range AF-III and AF-IV sets would be patterned after the two-way police radio sets produced by the Galvin Corporation and successfully tested in field maneuvers the month before. Though the OCSigO still felt that the "use of crystals for tactical purposes should be approached with the greatest caution," and warned Colton that the final decision to use these radios would be based on the possibility of future procurement of raw quartz and not on the engineering details, official policy had begun to shift.[41]

This did not mean that the OCSigO had abandoned its hope of decreasing or eliminating the dependence on quartz. In February 1941, Mitchell forwarded a letter to Colton describing a new technique proposed to eliminate the need for crystals. Colton responded that, though it was an interesting idea, "several other methods . . . now being developed . . . offer more promise."[42] In

April 1942, the Signal Corps was recommending to the Air Corps that crystal control be removed from some of their receivers. The Air Corps did not concur.[43] Development of master-oscillator sets requiring only one crystal was still under way as late as November 1941.[44]

By October 1940, the ARL (once in agreement with OCSigO in opposition to crystal control) was proposing to utilize it for the new UHF command set that was under development. They presented as evidence the opinions of "a number of leading radio development and manufacturing concerns" as to the necessity of crystal control. No other control circuit existed to meet the specifications of the radios, and industry representatives estimated a delay of one to two years in developing one. Research and Development endorsed the proposal but requested that the Procurement Planning Division give its assurance that sufficient supplies of quartz could be acquired to support both this Air Corps project and the Armored Forces sets.[45]

"THE PRESENT TREND SEEMS TO BE LEADING US INTO COMPLETE CRYSTAL CONTROL FOR FIELD EQUIPMENT"

As development continued at the Signal Corps Labs, the combat units continued to carry out field tests. A very dramatic test was carried out by armored units near Ft. Knox during the week of January 6, 1941. The tests utilized crystal-controlled SCR-245 sets mounted in scout cars (i.e., jeeps). Experiments were run by radio technicians to determine the ability of the sets to operate without interference from neighboring channels and the ease of setting up and maintaining radio nets. For the first test, four cars were utilized: one as the receiver, one transmitting the desired signal, and two transmitting interference signals at frequencies close to that of the desired signal. The vehicles were distributed over different distances with respect to the receiving car. It was found that, for a channel separation of 10 kHz, the desired signal could be received satisfactorily at a range of one mile, even if the interfering transmitter was as close as 50 yards. This range between transmitter and receiver could be increased to 5 miles if the channel separation between the desired and the interference signals was increased to 20 kHz (and 10 miles if, in addition, the interfering transmitter moved to 350 yards away from the receiver). A second test consisted of placing four cars side by side, transmitting simultaneously with separations of 10 and 8 kHz. A fifth car stationed 2.3 miles away attempted to receive the signals from the individual cars. This test was successful at both channel separations. A third experiment tested the abilities of the operators to make quick changes in frequency. It was found that the operators were able to switch their frequencies almost instantaneously. Lastly, a radio net was established using six cars separated by distances of up to 10 miles. In all tests, an average of only five minutes was needed to set up the networks.[46]

On January 20, 1941, the Armored Force Board communicated their results to General Chafee, to be forwarded to CSigO Mauborgne. Quoting from their report, these tests showed "that the use of crystals as a standard part of radio set SCR-245 results in a very definite improvement both in the method used and in the results obtained, in the frequency setting for net operation of each transmitter in the net." Furthermore, the Board found "that as a result of the standardization in frequency setting obtained with the use of crystals in the transmitter, better communication is obtained by less experienced operators than has heretofore been possible." The Board recommended to Chafee that all SCR-245 sets be furnished with complete sets of crystals.[47] Chafee forwarded the report to Mauborgne on February 3, and in his cover letter stated his approval of the report and stressed the Board's conclusions regarding frequency stability and ability to utilize more (narrower) channels within an allotted frequency band. He further requested that the groups that carried out the tests be allowed to keep the test crystals so as to continue their training.[48]

A very interesting report concerning these tests appears in the files of the Signal Corps. It is an unsigned and undated (though written sometime after March 15, 1941) draft of a report for the Executive Officer entitled "Crystals for Radio Set SCR-245." The report states that, though the set was designed to utilize crystals as well as a master oscillator circuit, relatively few sets had been supplied with their compliment of crystals. The fact that field testing of crystal control had been carried out is mentioned, with "reports indicat[ing] that all branches concerned are unanimously in favor of the issue of a full component of crystals." The report then described the benefits of crystal control with respect to allowing narrower frequency channels and rapid retuning. However, it also discussed six issues that it described as disadvantages. Of the more serious ones are the possibility that the SCR-245 might soon be replaced by the Armored Forces sets or the SCR-293 (based on the Link two-way radio set) and that the narrowing of transmitting channels would lead to a host of requests for the assignments of new channel frequencies. As with all previous OCSigO reports, the possible wartime shortage of raw quartz was pointed out. The final point made was a very important one: satisfactory use of a master oscillator–controlled radio required a fairly well-trained and experienced operator. If the using arms got into the habit during peacetime of using only crystals, the possible wartime supply of which was anything but certain, there might be a real shortage of sufficiently trained radio operators when the use of master oscillator–controlled radios again became a necessity.

The author made a very valid point. The entire question of crystal control was a gamble: should they stick with an established technology, in spite of its operational drawbacks and the wonderful promises of crystals, or should they bet everything on crystal control knowing full well that their procurement might become impossible during wartime? That decision had to be made soon. The report went on to point out that the using arms were obviously going to assume that these radios were going to be supplied with full sets of crystals, however, the Supply Division had not even placed crystals on its procurement schedules except in the cases of the field tests. The author urged the CSigO to make a final policy decision concerning not just the SCR-245 radio, but all Army radio. "The present trend," he wrote, "seems to be leading us into complete crystal control for field equipment."[49]

The unknown author of this report captured the essence of the problem facing the Signal Corps in the last year of peace before America's entry into the greatest armed conflict in its history. It *was* a gamble—which technology to pursue? In some sense, the decision had already been made by the combat arms; they had used crystal control, seen full well its advantages, and made up their minds that they had to have it. The die had essentially been cast in September 1940 with the approval of the Armored Forces series. This series of radio transmitters and receivers, however, represented only the initial trickle of the coming flood. By war's end, the vast majority of tactical radio would utilize crystal control, along with radio compasses and direction finders, navigation beacons, instrument landing systems, radio altimeters, and frequency meters (see Appendix 1). The stalwarts within the OCSigO were correct to keep stressing the question of supply, but they and everyone else only seemed to notice half of the problem. Even if a steady supply of raw quartz in amounts necessary to produce all of the needed oscillators could be maintained, *who would actually manufacture the units?* The unspoken assumption was that the current "industry" would simply expand operations and increase production (though with only a handful of companies producing a total of 100,000 crystal units a year utilizing essentially laboratory-scale techniques, the word "industry" is a bit too strong).[50] This belief was hopelessly shortsighted. The challenge before the Signal Corps, though unrealized at the time, would be to build, essentially from the ground up, an entire mass production industry. This job would soon fall primarily to Colton and O'Connell and the staffs of the Signal Corps laboratories and the General Development Section of the OCSigO.

3

THE SIGNAL CORPS LAYS
THE FOUNDATION

At 9 a.m. on Tuesday, July 11, 1944, in the North Ballroom of the Stevens Hotel[1] in downtown Chicago, a recording of the National Anthem was played. Those in uniform came to attention and saluted, while the civilians stood with hands over their hearts, many thinking at that moment of sons and brothers overseas. Assembled in the ballroom for the two-day conference on crystal production were representatives of the various Signal Corps laboratories and offices related to inspection, quality control, and specifications. Also present were representatives from nearly 150 crystal manufacturing companies;

Once the anthem had finished playing, Colonel L.J. Harris, Director of the Signal Corps Inspection Agency, and chairman of the conference that was about to get underway, strode to the podium to make the first of the morning's introductory remarks. He would be followed in turn by William Halligan and Paul Galvin, longtime radio manufacturers. Lastly, before the first coffee break of the morning, Major General Roger Colton, Chief of the Engineering and Technical Service (OCSigO) would address the audience.[2]

The truly remarkable aspect of the conference, apart from the vital information about to be conveyed to the manufacturers over the next two days, was the fact that such a gathering was able to take place at all. In 1940, when the decision to develop the crystal-controlled radios for the Armored Forces was made, no more than two dozen companies could be found in the entire country that played any role at all in the production of crystal oscillators. By the time of this conference, just two months short of the fourth anniversary

Crystal Clear: The Struggle for Reliable Communications Technology in World War II,
by Richard J. Thompson, Jr.
Copyright © 2007 by Institute of Electrical and Electronics Engineers

of the Armored Forces sets' approval, a full-fledged mass-production industry was in operation, producing over two million units *per month*. The creation of such an industry, from the most minimal of initial resources, was truly one of the phenomenal successes of the war effort. In the words of Bill Halligan that morning, the crystal industry was "truly a miracle of American ingenuity, inventiveness, and industry."[3]

This "miracle" was due to an extreme amount of individual hard work, but also to a remarkable spirit of cooperation—cooperation among private industry, the Signal Corps, and various government agencies. As Colton told the audience that morning, "The crystal industry is now one of the most cooperative of all the industries with which I have to deal. The teamwork between the crystal industry and the Signal Corps is splendid. Our decision to go into crystal controlled radios for widespread tactical use has been more than justified by the results obtained. The Army had radio before they had crystals. Now the Army has communications. That's the difference. Crystals gave us communications."[4] However, before Colton could speak those words of success in the summer of 1944, he had had to figure out how to create an industry out of the conditions in the summer of 1940. . . .

"A RECENT STUDY BY THIS OFFICE HAS DISCLOSED A DEFICIENCY OF PRODUCTION CAPACITY . . ."

From the beginning, the primary argument made against suggestions of converting to crystal control was the raw quartz supply. Already limited, it might actually be cut off during a time of war. No one, it seems, ever asked the more fundamental question: how the raw materials would be turned into finished crystal oscillators in the amounts that would be needed. This is partially due to the fact that, before Pearl Harbor, no one could have anticipated the magnitude of the need for crystal units that the war would bring. Furthermore, the crystal industry (and the Signal Corps procurement officers) seemed to be caught up in the same state of delusion or denial with regard to production capacity that affected most of the rest of the country's industrial sectors. Not only did most of the nation's industries believe their pre-war production capacity was sufficient to supply the wartime needs of the military, some, such as the steel industry, actually opposed expanding out of a fear of sinking a great deal of capital into increasing production and then having the market for this extra production fail to materialize.[5] Some of the larger pre-war crystal producers felt the same way, asking either for guarantees from the government that their increased output would be purchased or for government assistance in financing their plant expansions.[6] In July 1941, the Signal Corps Procurement Planning Section reported that the current manufacturing capacity for quartz oscillators had not been "overtaxed" and that only one or two facilities had needed to undergo any expansion.[7]

This perception, as with those regarding most sectors of American industry, was about to change. During the summer of 1941, a survey was commenced by the Signal Corps Procurement Planning Section to accurately determine the production capacity of the crystal industry. The manufacturers were asked both for their current production numbers along with an estimate of the amount of expansion they could hope to carry out. In making this expansion estimate, company owners were asked to consider the available labor pool, any needed machinery, the availability of raw materials, and the costs associated with such an expansion.[8] Along with established manufacturers, other firms that might be able to join the crystal oscillator industry were sought and queried as to their estimated production capacity.[9] Under the primary direction of Captain H.W. Zermuehlen of the Procurement Planning Division, the survey efforts continued throughout the remainder of 1941 and into early 1942. Though many companies were slow to respond to the survey letters, a much clearer picture of the industry as it stood at the time of the attack on Pearl Harbor did finally emerge. It was not a comforting picture, particularly with respect to the industry's ability to grind the needed quartz blanks and then to accurately finish them to the proper frequency.

A sense of urgency began to creep into Captain Zermuehlen's letters and reports by the end of the year. His discovery of a "deficiency of production capacity" became a common theme.[10] When potential manufacturers wrote asking for assurances that the need for oscillators really was increasing, Zermuehlen would refer to this deficiency and strongly suggest they contact the larger radio manufacturers for contracts.[11] He also wrote to companies, such as Western Electric, which had accepted contracts for very large numbers of oscillators (in some cases in the millions of units), pointed out this potential production problem, and asked just how they anticipated meeting their obligations.[12] On other occasions, he attempted to steer prospective manufacturers away from producing holders (which were not in short supply) and toward the grinding of quartz blanks.[13] It was becoming clear to the Signal Corps that something had to be done to augment the production capacity of the crystal industry; in fact, a real mass production industry needed to be created. Likewise, the Signal Corps realized that the job was too important to have the work scattered throughout several of its different branches. A new section devoted entirely to the production of quartz crystal oscillators was needed.

The abrupt start of the war did not catch the Signal Corps completely off guard. Its officers had been working for some time toward increasing its ability to quickly respond to the communication needs of the using arms, particularly the Air Corps. Though the ground forces (primarily the Armored Forces) had played just as important a role in the development of radio (and of crystal control), the needs of the Air Corps, through sheer weight of numbers, dominated the production and procurement efforts of the Signal Corps. Their requests for radios alone dwarfed those of all of the ground forces combined. These, coupled with their needs for radar and other communications equipment, presented a seemingly insurmountable task for the Signal Corps.

Furthermore, the Air Corps made it very plain that they expected the Signal Corps to support their needs above all others. In January 1941, a meeting was held between Air Corps Chief Henry H. "Hap" Arnold, Air Corps planners, and Signal Corps representatives. In response to a request from Arnold for a very large number of a particular type of radio, CSigO Mauborgne tried to point out the personnel difficulties such a request would cause as operators for these radios were almost as hard to come by as the radios themselves. It was quickly pointed out to Mauborgne by General George Brett, commander of the Air Corps Materiel Division and assistant to Arnold, that "he [Mauborgne] was a procurement man and not a tactician and that the Signal Corps should procure what the tactical people required."[14]

General Brett had never been a big supporter of the Signal Corps in general, or of Mauborgne in particular. The two of them had been involved in running battles over appropriations and expenditures for research labs and over control of airborne navigation equipment many times in the past.[15] The Air Corps' old complaints about the Signal Corps' "slow," "lethargic," and "unimaginative" R&D efforts began to resurface. With the dramatic increases in military appropriations for the 1940 and 1941 fiscal years, the Air Corps was very much interested in getting the most for their newfound money. Primarily through Colonel Alfred Marriner, Brett's anti-Signal Corps hatchet man, complaints to Chief of Staff General George C. Marshall intensified.[16] Marriner implored Marshall to do something to clean up the "chaotic mess" caused by the Signal Corps' apparent lack of support for Air Corps needs.[17] By summer 1941, Marshall decided he had to act. With Mauborgne due to retire in late September, Marshall ordered him to finish out the last six weeks of his term on a nationwide inspection trip and brought General Dawson Olmstead, Deputy CSigO and commandant of Ft. Monmouth, to Washington as acting CSigO.[18]

Olmstead went right to work, investigating the current development and procurement situations and designing organizations to address the problems. He also attempted to streamline the operations of the Signal Corps by grouping the various divisions into three branches, Administrative, Operations, and Materiel, thus replacing the current 11-man staff with a board of three branch directors. Placed in command of the Materiel Branch was Colonel Roger Colton. He would now be directly responsible for bringing to fruition all of the promises of crystal control that he and others had made over the previous five years.[19]

"SHROUDED IN THE MYSTICAL MISTS OF UTTER CONFUSION"

Throughout the fall and winter of 1941, numerous offices within the Signal Corps sought to clarify the quartz supply situation and determine the production capabilities of the oscillator industry. As each office had ill-defined responsibilities, a great deal of duplication of effort resulted. Such inefficiencies did not fit well within General Olmstead's philosophy of creating the proper organization for the job. The image of a single unit with sole responsibility for the quartz situation began to form in the minds of many Signal Corps officers. The first call for such an entity came from the Coordination and Equipment Division of the OCSigO. At a conference on February 13, 1942, in the old Munitions Building (a "temporary" structure built during WWI and still housing offices of the War Department), a recommendation emerged for a "centralized agency" to oversee and coordinate quartz oscillator production. Such an agency, the division suggested, should have authority over all aspects of production from the cutting of the raw quartz until the completed units were installed in their respective radio sets.[20]

One of the more chaotic aspects of the quartz program was the assigning of operating frequencies for the radio sets being shipped to the various field units. A great deal of confusion existed within the Signal Corps regarding which frequencies should be assigned to which units. This, along with the frequent changing of assignments after contracts for oscillators had been awarded, led to slow-downs and other problems for the manufacturers. A February 16, 1942, memo from Lieutenant Colonel J.D. O'Connell (currently heading General Development) to Colton described this problem in the context of producing radio sets for the Armored Forces. After describing the problem in detail, O'Connell called for a central agency to oversee all frequency assignments. In a draft of the memo, O'Connell showed his true feelings regarding the current situation. The original typed draft of the memo ended with the following sentence: "Without a centralized agency of this type, any semblance of order in the crystal field will remain a myth." In pencil, however, O'Connell marked out the last two words and added instead "shrouded in the mystical mists of utter confusion."[21]

O'Connell's call for a centralized quartz agency was soon answered. The Army Communication and Equipment Coordination Board met on February 28, 1942, and formally proposed to the CSigO that a Quartz Crystal Coordinating Section be created. Olmstead agreed with the recommendation and gave Colton, as Chief of the Materiel Branch, the responsibility for determining where within his branch it would reside. Colton gave it to O'Connell's General Development Division.[22] On March 5, 1942, Colton outlined the responsibilities of the new section:

    a. Take necessary steps to handle the crystal problem and successfully meet it.

    b. Determine the crystal frequency required for each contract.

    c. Determine the priority of delivery of crystals contracted and the allocation thereof.

d. Follow up on contractors and through them on subcontractors to insure crystal deliveries in step with set deliveries to avoid overload on any crystal producer.

e. Bring to the attention of all manufacturers of Signal Corps radio equipment existing Signal Corps Laboratories' studies on single crystal multi-channel and stabilized master oscillator systems so that any practical substitutions may be incorporated in production.

f. Require all manufacturers of Signal Corps radio equipment incorporating crystal control oscillators to complete by the earliest practicable date an engineering survey of the possibility of conversion of the design of each individual equipment to eliminate or to reduce its dependence upon quartz crystals, without significant degradation of performance below the quantitative limits imposed by current performance and specifications.[23]

It is interesting to note that, in the spirit of responsibilities "e" and "f" (first suggested at the Coordination and Equipment Division conference), the Quartz Crystal Section (QCS) was to work toward making itself unnecessary by dramatically decreasing or even eliminating crystal control. The old fears of dependence on crystal control were still around, and, in light of the production capacity surveys of the previous year, rightfully so.

Work began immediately on staffing the QCS. Initially, recruitment, at least of civilian scientists and engineers, was through university geology and physics departments and other scientific or professional organizations. The first scientists recruited by the QCS were William Parrish, Edwin Woods, and Maurice Druesne. Parrish, a mineralogist from Penn State, was assigned the problems of crystal orientation and production. He recruited two of his graduate school associates, Dick Stoiber and Clifford Frondel, to join the QCS in June 1942. Woods, a mechanical engineer recruited from the Patent Office, was involved with creating an equipment pool from which manufacturers could order needed pieces of production machinery. Druesne, a professor of physics at M.I.T. and former researcher at the Naval Research Laboratory, played the roles of field engineer and consultant. Along with these three scientists were six radio engineers, two secretaries, and three Signal Corps officers: Lieutenants Leslie Atlass, Charles Miller, and J.J. Lloyd. By the end of April, the QCS had added another physicist, Willie Doxey (later to be commissioned in the Signal Corps), two engineers, J.P. Randlett and E.N. Kagan, three more lieutenants, three secretaries, and a typist.[24]

One position the QCS lacked until June 1942 was an officer in charge. O'Connell played this role until Major Harry Olsen was placed in the position. Olsen was not new to the quartz project, however, having been involved peripherally since April.[25] By June, the QCS had reached more or less an equilibrium as far as staffing was concerned with a total of 38 personnel (15 of whom were scientists or engineers). Recruiting and staff changes continued throughout the duration of the war. Often the offer of a civil service position was enough to entice someone to join, but it was not uncommon to dangle the possibility of an officer's commission "for those hav[ing] the other required qualifications."[26] One such offer was made to Allyn Swinnerton, a geology professor at Antioch College in Yellow Springs, Ohio. Having already cleared a leave of absence with the president of Antioch, various members of the QCS (both scientists and high-ranking officers) contacted him in early 1943 regarding joining the group as an "expert consultant."[27] He joined the QCS in April 1943 in this position, but received an officer's commission to the rank of major in July.[28] By September, he was the Officer in Charge of the QCS, a rather rapid rise up the career ladder.[29]

Though the staffing situation had pretty much settled down by June 1942, just what exactly these people were supposed to do was still being worked out. Dick Stoiber remembered that summer as a "wild and wooly" time. "The crystal section was almost a completely independent project; an incoherent mess. We came and went as we pleased and for a while didn't even know who our boss was."[30] An inspection of the QCS files for 1942 leaves one with just such an impression. For the years 1940 and 1941, all of the materials from the Records of the CSigO related to quartz crystals are contained in only three file folders. The materials for 1942 fill 24 folders, with letters ranging from urgent Signal Corps appeals for crystal producers to requests from manufacturers for assistance with a wide assortment of crises. By the end of 1942, however, the letters and memos suggest a rather well-established industry, running fairly smoothly. However, a lot of work had to be done throughout 1942 in order to turn an "incoherent mess" into a smoothly running machine.

The very first project of the QCS was to collect, collate, and distribute all the known information related to the manufacture of quartz crystal oscillators. A great many manufacturers were entering the industry with no experience whatsoever in working with quartz; it was the job of the QCS to teach them. The members of the QCS had some learning to do on their own as well. This education was facilitated by gathering and reading all available publications regarding crystal oscillators and the properties of quartz, along with training visits to the established crystal manufacturers such as RCA and Bendix (Virgil Bottom, a physicist who joined the QCS in May 1943 remembers spending most of his time during his first few weeks on the job reading crystal oscillator production manuals).[31] The next step was to pass along this information to needy manufacturers. Letters announcing the existence of a QCS information pool were sent to manufacturers with offers of any help they might need (it was also made clear to the manufacturers that they should turn to the Signal Corps and no one else for help with their production problems).[32] By mid-April, the information pool was being hailed by the QCS as their first successful program.[33]

Along with gathering and disseminating what was already known about crystal unit manufacture, the QCS also had to keep abreast of new discoveries. This would require close cooperation between the QCS and the industry. Before any information gleaned from inspection visits or from shared information was distributed to the industry, the company from which the information originated was assured that they would be credited as the source of the information.[34] The standard cover letter accompanying any disbursements from the information pool contained a clause stating that the information had been provided by a particular company and that it was being furnished with the understanding that it would be "treated as confidential" and would only be utilized in the manufacture of crystal units under contract with the government.[35]

Of particular interest to the Signal Corps was the prevention of lawsuits pertaining to patent infringement. They hoped that clear identification of the originating manufacturer for any production techniques or pieces of equipment, along with this disclaimer statement, would protect the interests of the companies contributing to the pool.[36] A further inducement to contribute to the pool, for equipment manufacturers in particular, was the possibility of increased sales. Some companies, in addition to the gratitude of the Signal Corps for sharing information on their saws, drill presses, and grinding laps, were given mailing lists of all known quartz manufacturers.[37]

Sometimes information went out (with Signal Corps permission) directly from large manufacturers to needy quartz producers (particularly in cases where information on the use of a piece of equipment, such as the Western Electric X-ray orientation machines, was needed).[38] Other times a QCS advisor, while touring one plant, might happen upon a solution to a problem known to exist at another plant and be able to very quickly pass that information along.[39] By the fall of 1942, a series of newsletters, one from the SCL at Ft. Monmouth and the other from the Galvin Corporation, were being distributed within the quartz industry. The *Technical News Bulletins*, often one to two pages in length, contained information on production problems that had been brought to the attention of the laboratories, along with suggestions for increasing production and improving inspection results.[40] The other newsletter was likewise very short discussions of items related to quartz oscillator production originating from within the large system of companies subcontracting to the Galvin Corporation. More than 60 of these *Crystal Round-Table* publications (produced under the direction of Nick Anton) were distributed among the industry throughout the war.[41]

The highpoint of the QCS's efforts to disseminate information on crystal oscillator production was the publication in August 1942 of the *Handbook for the Manufacture of Quartz Oscillator Plates*. The *Handbook* covered information ranging from the procedures and equipment for the inspection of raw quartz and the proper determination of the crystal axes, to the cutting, grinding, and finishing to frequency of the oscillator blanks. By September, all 200 of the original printing had been distributed and a second printing of 300 was being planned.[42] Though more of a compilation of known techniques and information, the *Handbook* served the industry well until the publication of the much more professionally produced *Manual for the Manufacture of Quartz Oscillator Blanks* in February 1943.

Included with the *Handbook* were two posters designed to inspire factory workers toward increased production. One was a cartoon drawing of a U.S. soldier carrying as a spear a very long (and pointed) quartz crystal. The GI is using it to run Hitler, Mussolini, and Tojo over a cliff. In large letters across the top are the words "GIVE US THE CRYSTALS" and in smaller letters at the bottom left "AND WE'LL PUT THE . . . --- -. . .'s ON THE RUN!" (with "SOB" spelled out in Morse code). The other was a much less light-hearted cartoon. It showed an obviously dead solder lying face down in a trench, one hand on a radio set and the other lying next to the handset. Across the top of the poster are the words "HE LACKED A CRYSTAL!" and across the bottom "*IS IT* YOUR FAULT?" These posters were very popular with crystal producers; their supply being depleted as quickly as that of the *Handbook*. An additional order of 1,000 posters was placed with the Philadelphia Signal Depot within two months of the original printing.[43]

A companion project to the *Handbook* begun in late April 1942 was an industrial engineering study under the direction of Norman Kagan. Recruited into the QCS for just this purpose, the Signal Corps hoped that Kagan's study would be able to determine the best methods for mass producing crystal oscillators, information that could then be passed along to industry members. In cooperation with the Photographic Division of the Administrative Branch, camera and film analysis equipment was purchased by the QCS at a cost of $12,500.[41] Kagan, with a crew of Army photographers, began touring the production plants of the larger crystal manufacturers, along with some from other closely allied industries.[45] Ultimately, a "best practices" film was produced and used by field agents of the QCS as a training resource.[46] From time to time, Kagan's expertise was put to direct use as he served as an "on-site" consultant, helping plants overcome the numerous problems related to rapid expansion.[47]

One last resource that the QCS could count upon for support as it began to get itself and the crystal industry organized was the network of Signal Corps labs and signal depots. Work related to crystal oscillators had been going on for years at the General Development Laboratory at Ft. Monmouth, New Jersey, and the Aircraft Radio Laboratory at Wright Field, Ohio. With the concentration of all crystal-related work within the OCSigO, a similar gathering of crystal research under a single section took place at Ft. Monmouth (though the ARL continued its practice of distributing crystal research among its various radio development groups until early 1943).[48] In order to improve communication between the QCS and the Ft. Monmouth lab, an engineer was appointed to serve as liaison officer, spending half his work week at Ft. Monmouth and the other half at the QCS offices in Washington.[49]

With the 1940 decision to expand the military to 1.4 million men and the subsequent initiation of the peacetime draft, the military infrastructure began to feel the effects of rapid growth. The Signal Corps facilities were no exception. Ft. Monmouth, designated as a primary Signal training facility, began a pattern of expanding, filling to capacity, expanding, filling to capacity yet again, and further expansion.[50] During fiscal year 1941 alone, the Signal Corps labs grew by a factor of five compared with their 1940 size.[51]

By July 1942, the laboratories of the Ft. Monmouth Quartz Crystal Section had to be moved off the post entirely due to a lack of sufficient space. The labs took over an entire floor of a department store in Long Branch, New Jersey. The move was carried out with such swiftness that much of the stock of the former toy department was still in place. Thus, under the watchful eyes of baby dolls and rocking horses, the laboratory staff commenced their work. The Quartz Crystal Section would move its laboratories two more times before the end of the war; once to nearby Camp Coles Signal Lab (in April 1943) and ultimately to the Long Branch Signal Laboratory in September 1944.[52]

Whereas the Signal Corps labs would serve primarily as training and R&D facilities, the Signal Corps depots played the role of specialty production facilities. The largest of these signal depot crystal shops was at the Philadelphia Signal Corps Depot. This facility specialized in the production of crystal units that, for reasons of production difficulty, were not produced by private crystal manufacturers. During the summer of 1942, General Olmstead called for a second Signal Corps crystal plant to be established; this one to be housed at the Lexington Signal Depot in Lexington, Kentucky.[53] Severe shortages in crystal processing equipment delayed the opening of the plant. Though not expected to open until January 15, 1943, plans for staffing and training were drawn up during the fall of 1942.[54] However, by January 1943, with both the industry and the Philadelphia plant having greatly increased their output, General Colton, perhaps as a cost savings, decided to cancel plans for the Lexington depot.[55]

One other type of facility was developed at the Toms River Signal Laboratory in New Jersey. Though the primary responsibility for maintaining the supply of raw quartz for the crystal industry fell to the Miscellaneous Minerals Branch of the War Production Board (as will be discussed in detail in Chapters 6–8), the QCS was involved in both the procurement and quality inspection of raw quartz. The Raw Quartz Inspection Subsection was established at the Toms River Lab in October 1942 as a means of taking pressure off of the primary inspection facility at the National Bureau of Standards in Washington, DC.[56] In fact, throughout the summer and early fall of 1942, the QCS did whatever they could to help increase the rate of raw quartz inspection at the Bureau, including offering to loan the Bureau Signal Corps men from Ft. Monmouth, setting up their own inspection training program, and opening the Toms River facility.[57] Along with initiating its own inspection facility, the Signal Corps, under the direction of the QCS, carried out a development program at Ft. Monmouth for improved inspection equipment. Of highest priority was work done by Samuel Gordon on portable inspection apparatus.[58] These measures served as an important stop-gap during the early months of developing an inspection program at the National Bureau of Standards (see Chapter 6). Inspection wasn't the only involvement the early QCS had with raw quartz. It was also called upon by the industry and the War Production Board for help in expediting its delivery to the United States and also served as an intermediary between quartz importers and crystal manufacturers.[59]

"IT IS EVEN SAID THAT ONE PLUMBER CAME IN TO REPAIR A DRINKING FOUNTAIN AND WENT OUT WITH A CONTRACT FOR 150,000 UNITS"

Throughout the last nine months of 1942, the QCS struggled to define its mission and to learn how to carry it out. By the beginning of 1943, the QCS had become a very capable group of professionals, ready and able to help build the mass production industry needed to support military communications. Unfortunately, history did not play out in quite this sequence. From the first day of its existence, the QCS was expected to begin building and supporting that mass production industry; on-the-job training was the order of the day. Thus, while the right hand of the QCS was busy learning how to mass produce quartz oscillators, the left hand was just as busy finding the manufacturers with which to build an industry.

In a July 22, 1942, memo to the Chief of the Signal Supply Services, Major Olsen enumerated what he thought to be the three "major obstructions" to increasing the output of crystal oscillators: lack of manufacturers with the training to produce QCUs, lack of equipment for these manufacturers, and lack of funds for plant expansions, equipment purchases, and personnel training.[60] While all were true, the lack of trained manufacturers was the most serious. Before the attack on Pearl Harbor, the crystal manufacturers could be divided into two primary groups: small (sometimes one-man) outfits serving the needs of the amateur radio community, and the "Big Four." Of the group of small manufacturers, Bliley Electronics and Standard Piezo were perhaps the most established outfits, having been in business since the early 1930s. The "Big Four" consisted of Western Electric, RCA, Bendix, and Bausch & Lomb; large manufacturers of radio or optical equipment, maintaining quartz cutting facilities to support their own production lines (though Bausch & Lomb was primarily an optical instrument company, they were enlisted to cut blanks for crystal oscillators during the war).[61]

Due to their size and the amount of work they did with quartz, the Big Four companies all held positions on the pre-war National Defense Advisory Council's Quartz Crystal Committee. As this committee was charged with setting the standards for acceptable production techniques and with determining which types and sizes of raw quartz would be acceptable for the production of oscillators, these companies exerted a great deal of influence over the early industry. However, as the war progressed and the collection of smaller manufacturers grew into a full-fledged industry, these large corporations found it very difficult to keep pace with the evolution of machinery and techniques and ultimately became much less influential.[62] Still, due to its overwhelming size and the technical backing of its corporate sibling Bell Laboratories, Western Electric did manage to account for nearly 30% of the oscillators produced during the war.[63]

Initial work toward beefing up the existing manufacturers began within the Production Expediting Section of the Procurement Planning Division. Along with increased orders to the Big Four companies, efforts were being made to increase the production capabilities of some of the smaller plants (such as Bliley and Premier Crystal Laboratories). However, by February 1942, it was becoming clear that attempts to expand the current "industry" were not going to be enough. Offers for crystal contracts were coming back without bids, the established companies feeling unable to take on any more contracts.[64] This, on the heals of Zermuehlen's survey, along with worries beginning to surface as to the abilities of even such giants as Western Electric to produce their contracted amounts of oscillators, made finding additional manufacturers a very high priority for the Signal Corps.[65]

In addition to Zermuehlen's group, Signal Corps supply depots, procurement offices, and government agencies such as the Office of Production Management were enlisted by the OCSigO in the search for additional manufacturers.[66] Some manufacturers, however, seemed to doubt whether all the talk of promised large crystal orders could even be believed. Hesitant to initiate a costly expansion of his facilities, Theodore Valpey of Holliston, Massachusetts, owner of Valpey Crystals, wrote to the CSigO in February 1942 asking for reassurances that such an expansion was necessary.[67] Others, such as Earl Swanson of Apex Industries in Chicago, claimed to have heard in April 1942 that the crystal production crisis had already passed. He asked for guidance whether to enter the crystal business or not.[68]

By late April, however, the constant message that the Signal Corps still needed quartz manufacturers was beginning to bear fruit. A steady stream of letters, primarily from radio and electronics manufacturers, began coming in asking for details on converting to quartz oscillator production.[69] The Signal Corps directed some of the letter writers to contact established crystal plants in their area to learn more about the equipment and techniques required for crystal production.[70] Others were asked to contact the General Development Lab at Ft. Monmouth.[71]

Sometimes, in response to these letters, manufacturers were invited to meet directly with QCS personnel in Washington, DC, while still others came on their own accord.[72] During these visits, QCS personnel attempted to get a good idea of the production potential of the prospective manufacturers. Issues of plant size, finances, present operations, technical and engineering personnel, along with available production equipment and the personality of the owners were investigated. Those that seemed capable of converting to crystal production were usually directed to prime contractors within their geographical region for the possibility of subcontracting. Much less emphasis was placed on current manufacturing activities than on the ability to retool quickly. As such, a great many manufacturers who converted to crystal production had started out in such unrelated businesses as making lampshades, jewelry, refrigerated display cases, and tombstones.[73]

Virgil Bottom recalled that "anyone who even looked in the general direction of the Crystal Section was likely to get a contract." A common anecdote of the time concerned a plumber who came into the QCS office to repair a drinking fountain and left with a contract to produce 150,000 crystal units.[74] Nevertheless, many prospective quartz producers were turned down by the QCS. Many, lacking the necessary knowledge, experience, technical skill, and/or equipment, were strongly advised not to pursue financing for converting their plants to quartz production.[75] Other applicants, in addition to these deficiencies, engendered suspicions on the part of the Signal Corps due to their German ethnicity and recent contacts with German companies.[76] However, most of those advised against entering the quartz business were viewed simply as economic opportunists. One such applicant was described by Lieutenant J.J. Lloyd of the QCS as having "the over-confidence of a business executive who has not had to personally handle intimate production problems." Lloyd suspected that "someone had told [him] that crystal manufacture was an easy way to make big money, and with this in mind, he came to this office intending to walk-off with a big contract."[77]

In addition to face-to-face conferences, the facilities of some applicants were inspected by QCS personnel. These inspection trips gave the QCS personnel a much better feel for the potential of the manufacturer. Equipment could be inspected and those who would actually oversee production could be interviewed to determine the depth of their knowledge and experience.[78] Further, local business associates of the applicants could be interviewed with an eye toward determining the dependability of the plant owner in completing contracts on schedule.[79] As with the Washington interviews, many were approved for quartz production, but many were not.

Perhaps the most successful means of quickly developing a mass-production industry for quartz oscillators was through the subcontracting efforts of the established producers. By far the most successful of all the pre-war companies at setting up a network of subcontractors was the Galvin Corporation (later to be renamed Motorola) of Chicago. Galvin had been producing car radios since 1930, and seemed to the Signal Corps an obvious choice to manufacture their vehicular radios along with sets for artillery units and the Air Corps. Perhaps the most famous Galvin product was the SCR-536 "handy talkie" hand-held two-way radio. Common to all of these radios was the FT-243 quartz crystal unit. Galvin very early in 1942 found itself in need of hundreds of thousands of crystal oscillators for the radios they were manufacturing.[80] When Paul Galvin asked General Colton how he was supposed to carry this off, Colton told him, "Wait a minute, you [people] are industrialists; you are going to answer these questions. Go along as industrialists and take this thing and work it out yourselves and see how good industrialists you are."[81] At the time, Bliley Electronics was the primary crystal producer for Galvin. Paul Galvin put his chief engineer, Elmer Wavering, in charge of crystal procurement and sent him to Bliley to find out how soon they could carry out the 10-fold increase in their production needed to satisfy the Signal Corps contracts. Wavering was told it would take 10 years and a signed letter of guarantee from President Roosevelt before they could ever expand that much.[82] Needing crystals at a rate of over 13,000 per week, and being promised only 500 per week, Galvin had to find some other way to fill their crystal needs.[83]

One thing the Galvin Corporation did possess was a large staff of field representatives. Wavering sent his teams across the country with the charge to "find anyone who had some money, ingenuity, a real interest in winning the war" and talk them into producing crystal units.[84] At the July 1944 Chicago Conference, William Halligan (President of Hallicrafters, a rival radio manufacturing company) referred to Wavering and his men "sweeping across the country like a band of locusts," leaving very little business for Hallicrafters. Far from bitter about the loss of business, he went on to state that "Elmer and his crew did a very, very marvelous and very thorough job."[85] Ultimately, the network of Galvin subcontractors numbered 57 companies, spanning the country from Santa Monica, California, to West Orange, New Jersey.[86] At a March 1942 conference between the QCS and crystal manufacturers, O'Connell commented that "Galvin has subcontracted to almost everyone in the industry."[87] With the Galvin network comprising 43% of the industry members, this wasn't much of an exaggeration.[88] Though not the only radio manufacturer to utilize subcontractors (among others who did were RCA, Western Electric, General Electric, and Hallicrafters), Galvin became the indispensable man for the QCS. Companies writing to the QCS offices regarding entry into the quartz business were quite often steered toward the Galvin group.[89] Paul Galvin himself was in demand as a consultant and advisory board member.[90]

An immediate concern to those interested in retooling their plants for quartz production was financing. For those recruited into the Galvin network, one source of financial aid was the Galvin Corporation itself. Several subcontractors arranged advance payment on contracts to allow them to expand and to meet payroll and other operating expenses. From time to time, Galvin helped to secure loans for their subcontractors from both private banks and the federal government.[91] One source of Federal money in particular was the Reconstruction Finance Corporation (RFC). Created by an act of Congress in 1932, the RFC was charged with helping pull the country's businesses, banks, insurance companies, and other needy organizations out of the depths of the Great Depression. By mid-1941, the RFC had made loans totaling over $8 billion.[92]

In June of 1940, the RFC legislation was amended to increase its involvement in areas of national defense, specifically authorizing it to make loans "for plant construction, expansion and equipment, and working capital, to be used by the corporation in the manufacture of equipment and supplies necessary to the national defense."[93] Within a year, the RFC had made an additional $473 million in loans directly related to national defense.[94] The QCS played the primary role in certifying crystal manufacturers for RFC loans, with Galvin assisting in the cases of its subcontractors. After a satisfactory inspection by the QCS, a letter to the appropriate RFC office would be written on behalf of the manufacturer. The letters would routinely describe the type of work being done and its importance to the war effort, vouch for the company's ability to carry out the work, and reveal the monetary value of crystal contracts currently held by the company.[95] Frequent contacts between the QCS and the RFC offices might be required over the course of the loan application process.[96] In total, the crystal industry received $178,000 in RFC expansion loans.

Another source of government funding was the Defense Plant Corporation. This corporation, with a charge similar to the RFC, loaned eight crystal plants a total of $1.8 million (over $500,000 to Bliley Electronics alone).[97] Overall, the majority of the crystal plants needing financial assistance were able to secure it from their local banks or through subcontracting networks such as the Galvin group. Less than 10% of the industry acquired loans through the government.[98]

One of the most troubling impediments to the early expansion of the crystal industry was the shortage of processing equipment. As standardization in equipment did not exist at this time, crystal producers placed orders to equipment manufacturers under their own specifications. Thus, the equipment companies were faced with having to fill numerous orders for similar, but not identical pieces of equipment. In some cases, it just was not profitable for the equipment manufacturers to accept these orders.[99] A solution to this problem was worked out by the QCS with the financial backing of the Defense Supplies Corporation (DSC). The plan, developed during the summer of 1942, was for the QCS to create an equipment pool, similar to the information pool that had been in operation since the early spring. The QCS would place orders for large numbers of *identical* pieces of processing equipment that would be paid for by the DSC. Crystal manufacturers would then purchase their needed equipment directly from the DSC pool. This would enable the equipment manufacturers to easily produce the machinery in the amounts needed, as well as imposing a needed amount of standardization upon the fledgling crystal industry. In a June 24, 1942, letter to Ray Ellis, Chief of the Radio and Radar Branch of the War Production Board (having primary authority over quartz-related issues), Major Olsen of the QCS laid out the plan for the equipment pool and gave a list of the types of machinery that would be needed (e.g., saws, X-ray machines, and Polariscopes). The total estimated cost of the equipment came to $540,000. Olsen closed his letter with the following entreaty:

Since the Defense Supplies Corporation apparently was set up to meet just such emergencies as we are now confronted with in the crystal industry, we request that you proceed to enlist its whole-hearted cooperation in underwriting or guaranteeing payment to manufacturers of the various items of equipment which are urgently needed by the crystals manufacturers.[100]

To strengthen his chances of getting the DSC proposal accepted, Olsen sent a copy of this letter to the RFC, the parent organization of the DSC, emphasizing the importance of the request.[101]

In July, Olsen wrote again to the War Production Board, pointing out that his original cost estimates were too low. With improved estimates, and with the addition of a $150,000 contingency fund, the total request for DSC assistance came to $1.5 million.[102] This amended request was approved by the DSC and in early September, Ellis was able to inform General Colton that orders had been placed with 15 manufacturers for over $665,000 worth of equipment (with an addition $2 million in contracts for other, noncrystal related, Signal Corps equipment currently being negotiated).[103]

In July, Olsen had listed lack of manufacturers, equipment, and finances as the three "major obstructions" the QCS needed to overcome. By fall of 1942, they were well on their way to surmounting theses obstacles. One particular issue that continued to plague the industry throughout the war was a shortage of highly skilled workers to carry out the steps needed to turn raw quartz into electronic components. One interesting idea that was floated in late summer of 1942 was to utilize Japanese-American internees that had been evacuated from the Pacific Coast. The slow process of releasing the internees from the relocation camps had begun in the summer of 1942, with many of college age being released to attend school. Others were released to help with agricultural harvests in the western United States (though many were returned to the internment camps at the end of the harvest period). By October, fairly large numbers had been released and "relocated" to other regions of the country (primarily in the Midwest). For many of these former internees, the economic opportunities open to them were better than they had ever experienced on the West Coast. The War Relocation Authority, advised and assisted by civic and religious groups (such as the Quakers), attempted to assimilate the Japanese Americans into their new communities.[104] On August 25, 1942, Lieutenant Colonel James D. O'Connell was contacted by Melvin Scheidt, Acting Chief of the War Relocation Authority's Industrial Division about the possibility of employing "skilled artisans and craftsmen" in the Chicago or Kansas City crystal plants. O'Connell forwarded the letter to Olsen who then approached the Galvin Corporation and the Aircraft Accessories Corporation of Kansas City, Kansas. The offer, however, was subsequently turned down by both companies. No further efforts at placing Japanese-American workers in crystal plants were carried out.[105]

"ONE STEP TOWARD INCREASED CRYSTAL PRODUCTION WOULD BE THE EXCHANGE OF IDEAS AND PROCESSES BETWEEN AMERICAN AND BRITISH CRYSTAL PRODUCERS"

Along with assisting the nascent crystal industry in the United States, the QCS also cooperated closely with the British crystal manufacturing industry, both learning from and giving assistance to the somewhat more developed British industry. As a first step, Clifford Frondel was chosen to travel to England to learn what he could of the British manufacturing methods. After securing his draft release (a requirement for a civilian traveling overseas, even in the service of the U.S. government), and receiving the full range of immunization shots, Frondel left for England in early November 1942.[106] Frondel's four-month-long mission to England was very successful. Given complete access to all manufacturing facilities, he was able to carry out a very detailed survey of the production methods currently in use. This comparison of methods used by the two countries proved very helpful to the QCS. While there, Frondel also offered a great deal of assistance to the British. Working with their Quartz Advisory Committee (originally formed to focus on the problem of raw quartz supply), he helped to expand their efforts into the various areas of training and technical support currently being carried out by the QCS. Furthermore, a Technical Subcommittee was formed to act as a liaison between the research and development laboratories and the crystal manufacturers. Frondel also helped establish an information bulletin similar to those being produced in the United States, and authored 12 of them himself.[107] Perhaps of most assistance to the U.S. production industry was Frondel's study of British efforts to understand and control a very serious problem with crystal units known as "aging" (this inexplicable increase of an oscillator's natural frequency within weeks or months of production, coupled with a dramatic loss in vibrational "activity" is the topic of Chapters 9 and 10). Though he resigned from the QCS to take a research position with a major crystal manufacturer shortly after returning from England, QCS scientists would make use of Frondel's reports on this subject as they carried out their own research into the aging problem during the winter of 1943–1944.[108]

In return for the assistance and the hospitality shown to Frondel during his mission, the British were invited to send a representative for a similar tour of the American crystal industry. Captain C.F. Booth, a radio engineer with the Telecommunications Quartz Committee, arrived in the United States in June 1943. Booth was given a thorough tour of U.S. facilities, visiting over a dozen plants in the New England and Chicago areas.[109] The good relationship that developed between Booth and the QCS members continued throughout the war. This spirit of cooperation extended beyond issues of technical support. Millions of crystal units were provided to the British by U.S. companies on both lend-lease and cash-and-carry bases. Quite often, the QCS was called upon to play a troubleshooting or expediting role in securing these crystals in a timely manner.[110] The British were not the only allies to benefit from this type of QCS assistance. Poland, Russia, China, and Australia all benefited at one time or another from QCS assistance in getting their needed crystal units.[111] In addition to helping expedite orders from U.S. crystal plants, both Australia and New Zealand were given assistance in either setting up their own plants for manufacturing oscillators (New Zealand) or improving and expanding their current industry (Australia).[112]

One last example of QCS involvement in foreign affairs is a rather bizarre chain of events involving a neutral, nonallied country, quartz crystals, military intelligence, and concerns of Nazi industrial espionage. It began simply enough in 1941 with an order from the Swedish Legation for six crystal units. When the crystal units were ready for shipment, the manufacturer applied for an export license; the license request was denied on December 15, 1941. Upon being notified of this development, the Swedish Military Attaché, Karl Wessel, wrote the U.S. War Department asking for assistance in getting the export license approved. The War Department passed the issue along to O'Connell, who passed it on to Thomas Perrott of the QCS. In a handwritten note to Perrott, O'Connell asked him to investigate the issue: who made the crystal units? Can't the Swedes make these units themselves? Why only six? It all smelled "fishy" to him. It did to Perrott as well. He felt that these units might have been intended to serve as "educational guidance" for the Swedish crystal industry. It wasn't the Swedish industry that O'Connell was concerned about; it was the German one.[113] Though officially neutral, Sweden was surrounded by German-held territory and in a fairly precarious position; not occupied, but for the most part unable to deny any request of the Germans.[114] In his official report to the Chief of Military Intelligence, O'Connell stated his fear of allowing U.S. communications technology to fall into German hands. In his opinion, industrial espionage was the only explanation for the attempted purchase of these units; he recommended that the units not be shipped to Sweden but be sent to the QCS instead for further study.[115] The Swedish Legation complied with this request and the QCS gained possession of the crystals in June. One week later, in a two-sentence statement, O'Connell informed the Military Intelligence Foreign Liaison Branch that his office had examined the crystals, and after investigating their "manufacture and probable use ... withdraws all previous objection in this case."[116] No further explanation is given for the change of heart. It can only be assumed that the refusal of the export license (only one week after Pearl Harbor) and O'Connell's initial reaction were due in large part to the paranoia of a country still reeling from the brutal onset of war.

"THE CRYSTAL SECTION HAS WORKED FAR HARDER TO IMPROVE THE AIRCRAFT CRYSTAL SITUATION THAN ON ALL OTHER TYPES PUT TOGETHER"

Although a spirit of cooperation existed between the QCS and representatives of the crystal industries of allied nations (particularly the British), no such comfortable working relationship existed with the U.S. Army Air Corps. As discussed above, the Air Corps had long been dissatisfied with the support they received from the Signal Corps. On the issue of crystals for their airborne radio sets, their attitude was no different. Perhaps it was a case of an organization adopting the personality traits of its leader (in this case, the somewhat less-than-tactful General Hap Arnold), but the Air Corps, more often than not, played the role of the schoolyard bully when it came to fighting for industrial support for their communications needs. By November 1942, it seemed to an exasperated O'Connell that the Air Corps wanted nothing less than to have the entire production capacity of the crystal industry devoted to their needs.[117] Part of the problem was the difficulty in producing many of the crystal units needed for airborne sets. Due to the very strict specifications on these units, only a few companies were able to successfully manufacture them. This, combined with the extreme volume of crystals needed, inevitably led to delays in production. For example, by August 1942, production output of the DC-11-A crystal unit utilized in fighter planes was less than one third of the requested amount.[118]

In O'Connell's view, a large part of the problem could be tied to the detailed specifications drawn up by the Air Corps for their equipment; specifications, in his mind, written up without the proper benefit of technical expertise or production experience.[119] Many of the specification arguments made little sense to the QCS engineers who took part in their drafting. One instance dealing with operational ranges in temperature is particularly illustrative. Whereas the Army had set a temperature upper limit of 70°C (158°F), the Air Corps insisted that their crystal units operate at temperatures up to 90°C (194°F!). At a meeting concerning operating specifications, a representative from the Army asked "When that crystal unit in that airplane is at 90°C, where is the pilot and what is he doing?"[120] The Air Corps slowly gave ground on the specifications issues. With respect to operating temperature, they ultimately accepted the range used by British crystals, which had an upper limit of 30°C (86°F). Furthermore, when it was made clear to them that, with the exception of Bendix, no other company in the country would ever be able to manufacture the DC-11-A crystal unit under the current guidelines, they allowed for the substitution of a more easily manufactured unit. They also accepted a reorganization of airborne radio development, resulting in the creation of a Crystal Section at the Aircraft Radio Laboratory.[121]

Progress was made in improving the crystal outlook for the Air Corps, and in improving the relationship between the Air Corps and the Signal Corps, but at the cost of a lot of hard work. In a November memo to Colton, O'Connell bemoaned the apparent "feeling of stubborn uncompromise and concealed resentment" on the part of the Air Corps, while claiming the QCS to have worked "*FAR HARDER*" (emphasis in original) on Air Corps problems than on those of any other using branch.[122] The historical evidence does appear to back up O'Connell's claims. A great deal of effort was put forth in regards to design specifications, manufacturing capacity, and quality control of Air Corps crystals. By year's end, six of the industry's largest manufacturers (including

Western Electric, General Electric, Bendix, and Philco) were working exclusively on Air Corps orders. In addition, a dozen other smaller outfits were devoting most or all of their production to Air Corps crystals as well.[123]

By early 1943, production began to increase, due in part to an improved ordering procedure for Air Corps and British Royal Air Force crystals. One problem that had plagued production of aircraft crystals was the difficulty in working out the proper frequencies of the needed units. Oftentimes, orders would be held up in reaching the manufacturers while the proper frequencies were decided upon. The new procedure involved the placing of orders without initially specifying the frequencies. With at least a knowledge of the number of blanks that would be needed, a company could then begin procuring the necessary raw quartz and cutting the blanks. When the frequencies were ultimately determined and communicated to the manufacturers, the blanks could then be ground to their final form. This process of placing "blanket orders" for crystals greatly improved the entire process and, in particular, decreased the turn-around time for special orders of urgently needed frequencies.[124]

Still, the Air Corps kept their eye on the Signal Corps, and the crystal situation in general. In March 1943, Colonel Marriner, now Director of Communications for the Army Air Forces, wrote to the OCSigO, reminding him of the number of crystals needed during the current fiscal year and the number already received. As the difference in these two numbers amounted to almost two million crystal units, Marriner (politely) expressed hope that the needed production schedule would be met in order to complete the year's orders. His memo ended with a request for assurance from Olmstead that both raw quartz supply and production capacity would be maintained at a level sufficient to satisfy the needs of the Air Forces ("as well as those of other components of the Army").[125] Surely this admission by Marriner that "other components of the Army" had needs worth taking care of had to be taken as a positive development by the OCSigO. In response to this memo, Lieutenant Colonel Herbert Messer, now Officer in Charge of the QCS, gave assurances on both the raw quartz and plant capacity issues. In fact, Messer pointed out, the problem now being encountered with respect to plant capacity was not a lack of manufacturers, but the need to maintain production operations at "high efficiency" through the placing of sufficient crystal orders. The Air Forces could help themselves by actually beginning to order more crystal units.[126] This uneasy truce between the QCS and its largest customer would continue throughout most of 1943, though the calm would be shattered by the end of the year when the aging crisis peaked in intensity (see Chapter 9).

"COMMUNICATIONS BEING WHAT THEY WERE, THE MESSAGE WAS NOT RECEIVED"

Before taking over as Officer in Charge of the QCS, Colonel Messer was stationed in Rio de Janeiro as part of the U.S. Military Mission to Brazil. Old acquaintances, Messer and O'Connell kept in touch whenever they could. One such occasion took place in October 1942 when O'Connell wrote Messer a long, detailed letter regarding the crystal situation in the United States. In it, O'Connell commented: "We have had to do six months of damn hard work to get the manufacturing equipment of this industry developed."[127] In reading his description of the year's work, you can almost hear the exhaustion he must have been feeling. The previous year really had been a struggle. Though the problems related to Air Forces crystals were putting a real strain on the crystal industry, it did appear to be strong enough now to handle the pressure. That in itself was a remarkable accomplishment. From its beginning as a small group of overconfident large manufacturers and a rag-tag collection of small shops and one-man operations, the quartz crystal oscillator industry had grown, adapted, and developed into what could finally be described as a mass-production industry.

The products of this industry could not have been needed any more than they were at the end of 1942. American ground units were involved in large-scale combat operations for the first time in the war. On opposite sides of the globe, on the tiny island of Guadalcanal and the northern shores of the immense African continent, American soldiers and marines were finally taking their first steps on the long road to victory. Along with their training and conditioning, leadership skills, tactics, and weapons, the communications technology of the American GI would be severely tested in these regions over the coming months. Much to the chagrin of the Signal Corps, it appeared at times that the Allies' successes on Guadalcanal and in North Africa were not due to, but were actually in spite of, their communications technology.

The history of the North African campaign reads like a laundry list of communications failures. Brigadier General James Doolittle, serving with the 12th Air Force in North Africa, described the poor communications of both the air and ground forces as the "chief bugbear of efficient operations."[128] Communications between headquarters and the air fields being established early in the campaign were hampered by overburdened telephone systems. The airbases themselves lacked direction-finding, radio-ranging, and navigational beacon equipment, along with adequate radio equipment for air traffic control.[129] Contact with aircraft during operations was severely hampered, often jeopardizing entire missions. One such example was the airborne infantry raid on Tafaraoui, Algeria. When the airborne units took off from their airfields in southern Great Britain, their intelligence officers assured them that the Vichy-French forces holding Tafaraoui would not resist. This estimate changed, however, while the airborne forces were en route. Although attempts were made to alert the units to a possibly hostile reception, unfortunately, "communications being what they were, the message was not received." As it turned out, the mission was such a debacle that no planes even made it to Tafaraoui (most becoming lost and landing all across North Africa).[130]

A second example of a narrowly averted tragedy came at Oran, Algeria. An armored column of French Foreign Legion troops was spotted approaching the Allied-held airbase at Oran. Expecting attack from these troops, fighter planes were scrambled to attack the column. Again, intelligence estimates changed after the airborne attack had been launched. It was discovered that the troops, after having adequately defended their French honor by initially resisting the Allied invaders, were now on their way to surrender and join the struggle against the Axis. Contact was ultimately made with the fighter planes, just in time to cancel the attack.[131]

Complaints from high-ranking Air Force officers quickly made their way back toward Washington. In early December 1942, Major General Carl "Tooey" Spaatz, Allied Air Forces commander in North Africa, wrote to General Eisenhower, mentioning the "usual poor communications."[132] In a January 1943 letter to General Arnold, General Ira Eaker described "the breakdown of supply and communications" in Tunisia. "It is perfectly obvious that communications are at least ten years behind weapons. One of the severest handicaps to successful operations in North Africa is the feeble and inept communications system due to the inefficient or lack of development of proper reliable, secure, and efficient communications channels."[133]

The air forces were not the only participants to suffer from poor communications. The official U.S. Army history *U.S. Army in WWII: Mediterranean Theater of Operations: NW Africa: Seizing the Initiative in the West* has an extensive entry in its index under the heading "Communications Failure." A great deal of the ground force's equipment was lost or damaged during the amphibious landings on the coasts of Morocco and Algeria. Replacement equipment was slow in coming. Throughout the above-mentioned volume, frequent references are made to such things as "seriously inadequate" signal communications,[134] "marked deficiency in signal communications,"[135] and being "unable to establish radio communications with HQ."[136] At the Battle of Kasserine Pass, a costly defeat for the Allies, the loss of communication among the tank units and between these units and Colonel Robert Stark's headquarters seriously affected the Allies' ability to resist the German attack led by Field Marshall Irwin Rommel.[137]

The Allies did not have a monopoly on communications failures, however, with similar breakdowns affecting the German and Italian forces during the campaign.[138] And, fortunately for the Allies, sometimes things did work when needed. One particular instance that demonstrates both the wisdom of the 1939–1940 decisions regarding FM radios for armored units as well as the benefits of "Yankee ingenuity," actually started out as just another communications failure. The signal plan for the landings at Safi, French Morocco, called for a system of radio nets (systems of radios all operating on the same frequency) to be set up utilizing AM radio sets. These sets were not crystal controlled and, as such, proved impossible to keep tuned to the proper frequencies. Armored units within the area used their crystal-controlled FM sets to create a working radio net, efficiently handling both ground fire and air support requests during the initial days of the invasion.[139]

The blame for the communications problems of the North African campaign cannot be laid completely on the design of the equipment. A great deal of equipment was damaged during the landings. Other radios were lost to damage done by torrential rains.[140] Still, in other cases, communications equipment never reached its intended unit due to failures in Signal Corps procurement and distribution.[141] Some problems were due to poorly trained Signal units, many of which had been formed just prior to joining the invasion fleet on the U.S. East Coast. Some received the only training they would get on their particular equipment while crossing the Atlantic Ocean.[142]

One instance that did suggest a failure of design, with potentially ominous consequences for the crystal industry, is described in the U.S. Army history. Shortly after landing in North Africa, a tank unit found its radio sets to be inoperable. The history describes the equipment as having "been put out of order by the long period of disuse while en route by sea."[143] This sounds very much like a failure of the crystal units due to aging, most likely accelerated by the humid conditions of the ocean voyage. Similar inexplicable radio failures were being experienced in the likewise humid environment of Guadalcanal. Sergeant Roy Smith, a communications line chief with the 68th Fighter Squadron, remembered problems with crystals. "Sometimes, we'd lose entire conversations with the pilots. After a while, they'd come back in and report problems with the radios. We'd check the equipment and it was fine"[144] Ground forces also suffered humidity-related radio failures. Electronic equipment suffered a great deal of corrosion, leading to broken circuits and electrical contacts. The extreme moisture led to short circuits and the altering of frequencies of crystal units.[145]

In the absence of reliable radio communications, the ground units on Guadalcanal resorted to the old standby—wire. However, the potential drawbacks of wire-based communications, first hinted at in the pre-war maneuvers in Louisiana, were experienced fully on the island. Advancing units frequently outran their wire communications, forcing rear echelon and HQ units to continuously relocate to maintain contact.[146] Even the vaunted security of wire communications was demonstrated to be much less than earlier believed. On many occasions, English-speaking Japanese troops managed to tap into the telephone lines and, impersonating American GIs, request assistance. The responding patrols were subsequently ambushed.[147] This ploy was used by the Japanese throughout the war in the Pacific.

Overall, the initial showing of Signal Corps equipment and personnel was substantially less than had been hoped. Training, procurement, distribution, and, to a certain extent, design and manufacturing problems led to tremendous difficulties in both North Africa and Guadalcanal. The communications industry, including the crystal manufacturers, would respond to the shortcomings of their equipment in these first campaigns. In a sense, the year 1942 was spent just figuring out how to produce communications equipment in the amounts needed. Quantity overshadowed quality. This was particularly true in the crystal industry.[148] Valuable lessons were learned in these campaigns; lessons that would be applied in the development laboratories and on the production floors of the American radio manufacturers and, under the guidance of the Quartz Crystal Section, in the crystal manufacturing plants as well.

A great deal of the credit for the 1942 efforts toward developing the crystal industry belongs to Colonel O'Connell and his group of scientists, engineers, expediters, and procurement specialists. However, as will be discussed in the next chapter, an equal amount of credit belongs to the industrialists, small plant owners, out-of-work civilian-radio merchants, and self-taught crystal grinders who actually composed the industry. The development of the quartz industry was far from complete; it would go through several more crises and evolutionary advances over the coming months. The QCS might have laid a foundation for an industry in 1942, but now, in 1943, that industry had to be supported (with information, training, equipment, financial assistance, and a steady supply of raw quartz) if it was going to achieve the level of quality production so obviously needed by the American military.

4

NOTHING ELSE TO DO BUT GRIND CRYSTALS

Louis Patla started his radio coil business in 1933 with five dollars in cash and a coil winding machine he'd constructed from an old sewing machine found abandoned in an alley. Employed in the test lab of a Chicago radio manufacturer during the day, the newlywed would spend his evenings in his basement making the radio tuning components. His partner Morris McLean worked the night shift at the same company and acted as the salesman and business manager for their sideline business during the daytime. The business grew slowly at first. Their invention of a new type of "loop aerial" antenna, however, turned the sideline activity into a full-time operation. By 1940, DX Radio Products was a very successful company employing nearly 250 people. Then the Japanese bombed Pearl Harbor.

The very next day, the Federal Communications Commission forbade the operation of all amateur radio stations. Soon after that, as with many other consumer goods, the production of civilian radio products was halted. DX Radio Products, along with most of the rest of the smaller radio manufacturers across the country, suddenly found itself out of business. Companies with outstanding orders were allowed to complete production on those items. As their last production run was nearing completion, Patla and his chief salesman, George Timmings, went across the street to the local bar to see if the change of scenery could help them figure out what to do next; other than joining the Army, their options seemed very limited. During their conversation, Patla mentioned offhand that, earlier, he had been thinking about

Crystal Clear: The Struggle for Reliable Communications Technology in World War II, by Richard J. Thompson, Jr.
Copyright © 2007 by Institute of Electrical and Electronics Engineers

expanding into the manufacturing of crystal oscillators. An amateur radio operator himself, Patla had read articles in *QST Magazine* about these frequency control devices. Timmings suddenly became very interested. Could Patla really manufacture these devices? Patla thought that he could; his opinion was that, if it were possible for anybody to manufacture them, surely his group of "smart boys" could do it. Timmings quickly paid their tab, mentioned something about having business to take care of at Galvin, and hurried out of the bar. Later that afternoon, an obviously pleased Timmings returned to the DX plant. Calling Patla and McLean into his office, he showed them the product of his afternoon's work: an order from the Galvin Corporation for 10,000 DC-5 crystal units at a price of five dollars each. The DX Crystal Company, the first crystal manufacturer in Chicago, was born (and two men were immediately dispatched to the public library to find those *QST* articles).[1]

The story of Louis Patla and DX Crystal is not that different from any other crystal company born of the war. All across the country, men who had managed to build successful businesses during the Depression years now found themselves out of business. Many of the more technically savvy manufacturers answered the call for crystal oscillators. As the story of DX Crystal Company shows, getting orders was relatively easy; figuring out how to manufacture the units to fill those orders was much more difficult. In fact, other than the pressures of the war, the obstacles facing the new crystal manufacturers were little different from those faced a decade earlier by the pioneers of the industry.

"CUTTING AND POLISHING PETRIFIED WOOD FOR CHESS FIGURES WAS ONE OF MY FATHER'S HOBBIES"

If any one place can be called the birthplace of the crystal industry, Carlisle, Pennsylvania, is the leading contender. Known to its residents as "Crystal Town, USA," Carlisle was home to some of the most successful pre-war crystal manufacturers. All of these early companies can be traced to a single man, Grover Hunt, and a single institution, Dickinson College.[2]

At Dickinson College, at the beginning of the 1930s, three important ingredients came together: enthusiastic students with an interest in amateur radio, an inspiring professor, and an experienced machinist with an entrepreneurial drive. Dr. W.A. Parlin was hired to chair the physics department at Dickinson College in 1930. Upon his arrival, he discovered the amateur radio interest of three of his physics students and helped them to set up the first Dickinson radio station. The students, Edward Minnich, Howard Bair, and Charles Fagan, soon learned of crystal units and, with the financial backing of Minnich's father, acquired one for their station. The desire to build their own equipment seems to be a standard part of the character of a ham radio enthusiast. The Dickinson students, being no different, next decided to attempt to manufacture their own crystal oscillators. Under the tutelage of Dr. Parlin, the students learned to orient, saw, and grind their own crystals.

At the same time, Grover Hunt, a general maintenance man for the college, was involved in a similar pursuit. Working at Dickinson was a family affair for Hunt. His wife was the campus nurse and his father-in-law was a night watchman. Hunt was a chess enthusiast and had long desired to build his own chess set. During a cross-country vacation trip in early 1930, he collected a quantity of petrified wood, which he hoped to cut and polish into a set of chess pieces. With the help of his brother-in-law, P.R. Hoffman, a trained machinist, Hunt began working on his hobby in the basement of the college science building. Spending his evenings in the machine shop, Hunt came to know the students involved with the radio station and to learn of their work making crystal oscillators. Whereas the students viewed their work as an interesting hobby, Hunt saw the potential for a successful business. Joining forces, Hunt quickly learned to produce quality crystal blanks while the students became his sales force.[3] This partnership represents the first commercial crystal operation in Carlisle.

By late 1934, Hunt's one-man business was becoming well known throughout the region. One of his biggest customers was the Goodyear Blimp company in Akron, Ohio. In early 1935, its chief engineer, Linwood Gagne, stumbled upon an opportunity to enter the crystal business in a big way. Another of Goodyear's crystal suppliers had gone bankrupt, leaving several large government orders unfilled. The local bank now in possession of the company's equipment and responsible for the contracts offered to sell the entire concern to Gagne. Gagne approached Hunt with the idea of forming a partnership; Hunt approved and the Standard Piezo Company was founded. With Gagne's experience in radio engineering, Hunt's experience with crystal manufacturing, and a great deal of help from Hunt's machinist brother-in-law, Hoffman, Standard Piezo became a huge success. During the war, it would become one of the largest crystal producers in the country, employing nearly 1,200 people at its plants in Carlisle and Scranton, Pennsylvania.[4] Furthermore, it was the first crystal plant to reach a production rate of 1,000 units per day and to receive the Army-Navy "E" award for excellent service in support of the war effort.[5] Grover Hunt would not be a part of the wartime Standard Piezo; he and Gagne had a parting of the ways within a year or so of their partnership. Hunt left Standard Piezo to set up a competing company, G.C. Hunt & Sons, in his father-in-law's garage. Hoffman, on the other hand, now had *two* local customers for his machine shop. With the coming of war, Hoffman himself entered the crystal business, operating a finishing plant.[6] One other plant, Carlisle Crystal, was formed during the war by a retired Signal Corps officer, Colonel Philip Mathews, but did not survive long after war's end. Standard Piezo, G.C. Hunt & Sons, and P.R. Hoffman, all "descendents" of the early radio work at Dickinson College, were valuable crystal producers during the war and served as the nucleus for the post-war industry, going through numerous mergers and changes of ownership.[7]

At about the same time as Grover Hunt and the Dickinson students were learning to produce quartz crystal oscillators in Carlisle, another Pennsylvania company was getting started in Erie. F. Dawson Bliley had received his first amateur radio license at the age of 14. From then on, much to the consternation of his mother, his home was the scene of much radio experimentation and development. Learning to cut and grind quartz while in college, Bliley soon became known among the hams of his area; starting out making crystals for his friends, he eventually began selling his units. In 1930, the Bliley Electric Company was formed and housed in the basement of his home. Soon Bliley had engaged a couple of ham friends to help him produce crystals. Later, a salesman was added to the company.[8]

By 1931, Bliley's company had expanded beyond the confines of his mother's basement. A partnership was formed with a local optometrist, which gave Bliley access to motor-driven grinders, greatly speeding up the manufacturing process. From there, Bliley grew into a regional supplier of crystal oscillators. Known as much for his radio skills as for his company's products, Bliley was recruited to accompany Admiral Byrd on his 1933 trek to the South Pole. Though he did not accept the invitation, he did donate 100 Bliley crystals for the Admiral's use (and did not fail to mention this in his company's advertisements). The Bliley Electric Company was known for innovation, in both its products and its manufacturing procedures. One of its most important manufacturing innovations was the introduction of acid etching into the production of crystal oscillators. Using acid to etch the final layers of quartz from the blank gave a dramatic improvement in crystal performance. The importance of this innovation, and Bliley's attempt to keep it a "trade secret," will be discussed in later chapters.[9]

After a decade in the crystal business, Bliley had become one of the most successful companies in the industry. As such, it was in a prime position to be of immediate help to the Signal Corps during the scramble to procure adequate numbers of crystal units during the final months of peace and the early months of war. Even before the war's commencement, Bliley had been working closely with the SCL in designing and testing crystal units for Signal Corps equipment. As early as April 1941, Bliley was in contact with the OCSigO regarding orders for large numbers of crystal units. Knowing that the OCSigO was "deeply concerned" about the crystal situation, they wanted to help insure an "adequate supply of quartz crystals."[10] This early work on the part of the Bliley Company was a great help to the Signal Corps once the country was finally drawn into the war.

"I DIDN'T WANT TO LET WWII GO BY WITHOUT HAVING SOME PART IN IT"

Perhaps the company to make the most rapid, efficient, and profitable transition to wartime production was the Monitor Piezo Company of Los Angeles (now located in Ocean Side, California). Formed in 1928 by Herbert Blasier, a radio ham since 1912, Monitor's initial business consisted of producing crystals for ham radio operators, commercial stations, and fire and police two-way radio sets. Up until the attack on Pearl Harbor, Monitor Piezo was a sideline business for Blasier, who was employed as an electrical engineer for the Southern California Telephone Company.[11] With the onset of war, though, Blasier made the crystal business his sole occupation. The planning for his wartime expansion was very meticulous. First, Blasier made a survey of the quartz importers, endeavoring to find out which type of raw quartz was least popular with other manufacturers, but still available in large quantities. He then designed his production methods to utilize this and only this type of quartz (small, faced crystals known as "candles"). Having greatly decreased the possibility of competition for raw quartz, Blasier then set about locating a site for his expanded facility. He settled on a "moderately prosperous" residential section of Pasadena and found there a good supply of labor, primarily young female high school graduates. In hiring, Blasier stressed that the girls be intelligent, educated, and, if possible, have a relative in the armed services. These characteristics, combined with good wages, insured a hard-working, dependable workforce.[12]

Not all of the Monitor employees came from the immediate neighborhood. One in particular came from the west Texas town of Abilene. Edith Lineweaver, after months of watching the boys from her high school go off to the military, and many of the girls go off to work in industry, decided to leave home herself. After responding to a newspaper ad for training in Dallas and guaranteed West Coast jobs, Miss Lineweaver and a friend found themselves working as riveters in a San Diego aircraft plant making B-24 bombers. After about a year of this work, she heard about job opportunities in Pasadena and was hired to work at Monitor.[13] The work at Monitor was definitely cleaner, safer, and less exhausting than the work in San Diego had been. Along with the available recreation area, on-duty maid, and free coffee, the plant workers were trusted to complete their duties without the pressure of a plant foreman or forewoman. This "very well thought out plan" (in the words of a Signal Corps inspector) led to an extremely low employee turnover rate coupled with a very low rejection rate for finished oscillators (0.4%).[14] Though Blasier's commitment to building his production line around a single type of raw quartz would lead to problems of supply at various times throughout the war (see Chapter 6), the well-planned expansion of his business was a model for others to emulate.

Along with the larger established companies, the pre-war industry included several small operations, many of which were one- or two-man operations. Many, like Bliley and Patla, were ham radio hobbyists who decided to make money producing crystal units. A.E. Miller, of North Bergen, New Jersey, was quite possibly the very first commercial producer of quartz oscillators. A trained lens grinder, Miller left the optical industry in 1923 to begin selling crystal units to hams.[15] Though he started much earlier than Bliley or Hunt, his business never attained the size or prestige that these companies did. In fact, Miller's entrepreneurial "zeal" often brought him to cross purposes with the Signal Corps during the war (see Chapter 7).[16]

Another small operator whose work had far-reaching effects during the war was Herb Hollister. Beginning in 1927 at the suggestion of a ham friend, Hollister began making crystal units for Collins Radio Company (owned by Arthur Collins, another ham friend).[17] Twelve years later, Hollister was living in Wichita, Kansas, running a small AM radio station, and still producing crystal units in his basement. His business expanded somewhat, however, upon receiving a contract from Galvin to produce crystals for their new two-way radio sets. Wichita is home to Friends University, and Hollister made use of the local student labor pool for his expanding business. Some of his student workers were physics majors, being taught by a young professor named Virgil Bottom. Through his students, Bottom became friends with Hollister and, at the same time, developed an interest in quartz oscillators. It was in Hollister's shop that Bottom learned the physics, along with the production methods, of crystal oscillators that led to his being recruited into the QCS in 1943.[18] Hollister moved his business to Boulder, Colorado, in 1942, expanding to about 20 full-time employees.[19] Bottom would take a position at Colorado A&M University in Fort Collins, Colorado, in the fall of 1942 and continue their collaboration.

When asked in 1981 about his reasons for participating in the wartime crystal industry, Hollister wrote: "As to my reasons for expanding during the war, I am quite hazy about that too. However, I am sure that I was partly influenced by the pure fascination of working with crystals and developing new methods of production. Also, there was the feeling that I didn't want to let WWII go by without having some part in it."[20] Though his business never rivaled Bliley or Standard Piezo, Hollister's company (and dozens of others like it) played a vital role in the crystal industry. Perhaps his most important contribution was in providing Virgil Bottom the opportunity to learn about crystal oscillators. Upon joining the QCS in 1943, Bottom was the only member of the technical staff to have any first-hand experience with the manufacture of oscillators.[21] This, combined with his training as a physicist and his experience as a teacher, served the Signal Corps and the crystal industry extremely well throughout the remainder of the war (see Chapters 5 and 10).

"CLEARLY I HAD TO LOOK FOR SOME OTHER WAY TO SURVIVE THE WAR YEARS"

By the end of 1942, crystal producers that had been in business before the war were very much in the minority. A great many, like Louis Patla and Morris McLean, had been in the commercial radio business. Converting their plants to crystal production was an act of survival. This was the case for Leo

Meyerson. Having built radios since the age of 13, when he also discovered, to his embarrassment, that it was illegal to broadcast without a license, he started his own business in 1934 with a $1,000 loan from his father. He had to promise to come to work in his father's grocery store if the radio enterprise was not a success. By 1940, his business was "getting into high gear." After the war started, and his radio business ended, Meyerson heard through the grapevine that "the military was looking for quartz crystals and that they were having some difficulty finding them." Knowing nothing of crystal production himself, he called an old ham friend who did. As both needed some "way to survive the war years," they decided to investigate this business possibility.[22]

With a "test order" for five crystals from Galvin, Meyerson and Al Shideler went to work. For three days and nights they worked at producing the crystal blanks. An all-night bus trip from Council Bluffs, Iowa, brought them to the office of Dan Noble, Chief Crystal Engineer for the Galvin Corporation. Noble took the crystal blanks for testing and left Meyerson and Shideler waiting for several hours. Eventually, he returned with the news that the crystals had passed the tests. "Do you fellas think you can handle an order?" he asked. "We can handle anything you can give us," was Meyerson's reply. Meyerson and Shideler caught the bus back to Council Bluffs carrying an order for 80,000 crystals at $8.53 each. Scientific Radio Products, formed immediately after their return from Chicago, eventually employed 500 workers and produced an average of 25,000 crystal units per month.[23]

Not all radio component manufacturers were put out of business by the war. Those that were large enough to supply components to prime Signal Corps contractors in the amounts needed continued their operations. Some also expanded their interests into the quartz oscillator field. One such company was the Good-All Electric Company of Ogallala, Nebraska. Trained as a jeweler, Robert Goodall began to "tinker" with electrical apparatus for jewelry-related jobs such as soldering and watch cleaning. This ultimately led to a full-time business producing a wide variety of electric products, including a great deal of radio-related items. Often described as a "green horn," Goodall had a knack for entering a business with no prior knowledge and rapidly becoming a success. Such was his experience with crystal oscillators. Virgil Bottom remembered him from an interview as having a "great deal of enthusiasm," but little knowledge of crystal oscillators.[24] The Good-All Company began making crystal oscillators shortly after the war began and continued throughout its duration.

Another midwestern radio company that expanded into crystal production was the Aircraft Accessories Corporation of Kansas City, Kansas. A pre-war producer of aircraft radio equipment, Aircraft Accessories was hoped by the Signal Corps to become the nucleus of a midwestern crystal industry.[25] Though such an industry did develop, Aircraft Accessories did not exactly play the role the Signal Corps had hoped. The company was having serious difficulties getting into production at the time of the Pearl Harbor attack. Though the company blamed their problems on a lack of skilled labor, Signal Corps inspectors believed the problem lay instead with management.[26] Seven months later, Signal Corps inspectors were still reporting serious problems with the plant. After a visit to the company, QCS members Lloyd and Frondel reported that they still didn't seem to know what they were doing with respect to cutting quartz.[27]

John Ziegler, a crystal engineer from RCA, was sent to Kansas City as a Signal Corps consultant to see what could be done about the situation there. His conclusion was to forego attempts to straighten out Aircraft Accessories and to start his own company. Ziegler convinced George McGrew, a treasurer at the nearby North American bomber plant, to go in with him on the venture. Their company, Crystal Products, Inc., became very successful and ultimately played the role that had been hoped for by the Signal Corps for Aircraft Accessories. In a similar sense to the Carlisle industry, most of the later wartime and post-war crystal companies in Kansas City could be traced to Ziegler, McGrew, and Crystal Products.[28]

Perhaps Ziegler's biggest contribution to the development of a Kansas City crystal industry was a course on piezoelectricity that he taught at the Kansas City Junior College. Using patent applications released by Bell Labs as textbooks, Ziegler started his course with 20 students (some of whom made lifelong careers of the crystal business). One such student was Ernest Ruff. Ruff was hired as the first employee of Crystal Products (ostensibly because his carpentry skills were needed for building the workbenches of the plant). Trained initially as a crystal finisher, Ruff learned quickly all aspects of production and ultimately became the chief troubleshooter of the company.[29] Eventually, the Kansas City area became home to 10 crystal companies.[30] Several were started at the suggestion of Ziegler and benefited from the troubleshooting expertise of Ruff.[31]

With the exception of the handful of companies that existed prior to the war, most of the wartime crystal producers were either brand new companies, sideline operations of larger radio producers, or refugees from some other business arena. Of the latter group, many were like Leo Meyerson and Louis Patla, small radio component manufacturers put out of business by wartime restrictions on the amateur radio community. Others, however, had no connection to radio or electronics at all, making such things as sound recording equipment, refrigerated butcher's cases, camera accessories, and, in the case of Irving Lieberman of Chicago, lampshades. What all of these converts and refugees had in common were owners and operators who, having built up companies through the dark days of the Depression, refused to see them end just when the economy seemed to be getting back on track. Patriotism and a desire to contribute to the war effort played a role, but the heart of the issue was economic survival. In the words of Meyerson, they were all looking for a way to survive.

"THE PLANT IS LOCATED IN A CONVERTED CHICKEN HATCHERY THAT HOUSES THE EQUIPMENT, THE GREAT MAJORITY OF WHICH IS HOMEMADE"

One of the reasons these men had prospered during the Depression was their ingenuity and, in the words of Robert Goodall, the ability "to find a new use for an old idea."[32] Their original businesses had been built upon such foundations, and their crystal operations would be no different. To produce a crystal oscillator, a quartz rock, one of the hardest known substances, must be cut, ground, and lapped into a wafer only a fraction of inch thick. As there existed no commercially available "standard" pieces of equipment for carrying out such operations, and as most of the plant owners could not have afforded such equipment even if it were available, most newcomers to the crystal industry built their own equipment. The pre-war companies had all begun operation in this way and several still held fast to their homemade equipment. A July 1942 inspection of the Valpey Crystals plant in Holliston, Massachusetts, found "the great majority" of their cutting and grinding equipment to be homemade. Though not producing exceptional amounts of crystals, the QCS inspector was impressed that they were able to do as well as they did (about 700 crystals per month) with the equipment they were using.[33] As late as January 1943, the QCS was warning manufacturers of test equipment about "the tendency of crystal manufacturers to build their own equipment" and their "reluctance to purchase commercially available test equipment."[34]

For small converts and refugees like Louis Patla, designing and building their own equipment was an economic necessity. Patla himself was a master at redesign and adaptation of existing equipment. His first quartz lapping machine was a converted bread dough mixer. He later made use of the large inventory of home-workshop drill presses at the local Sears & Roebucks for building improved lapping and sawing machines. One particular piece of equipment employed in the proper orientation of raw quartz before cutting used polarized light. Polariscopes, as they were known, were commercially available, but were expensive and required permission from the government to purchase. Patla built a crude one for himself using a pair of Polaroid sunglasses purchased at the neighborhood Walgreen's drugstore. The drawings for the homemade polariscope were traded to quartz manufacturers Leon Faber and James Knights for their drawings for homemade drill-press lapping machines. The production of both companies increased dramatically as a result of this cooperation.[35]

Leon Faber was quite an innovator in his own right. Employed full-time by the Illinois Northern Utilities Company, Faber had been producing crystals with James Knights since 1932. Both ham radio enthusiasts, Faber and Knights had set up a crystal cutting and grinding operation in the back of Knights' gas station in Sandwich, Illinois.[36] As wartime shortages and rationing mounted, the crystal operation began to take over from the gas station as the primary source of income for Knights.[37] At the urging of the QCS, Faber was given a leave of absence from the utility company in order to run the crystal operation full time.[38]

Most of their early sawing equipment was designed by Faber and built by a local blacksmith.[39] The innovation that brought him into contact with Louis Patla was the double-sided grinding machine. It was very important in the grinding of crystal blanks to keep both sides as close to parallel as possible. Grinding one side, inverting the blank, and grinding the other sometimes introduced differences in the two sides' orientation. Faber developed a machine that lapped both sides of the blank at the same time, guaranteeing the sides would be perfectly parallel.

Perhaps his most important homemade innovation dealt with the proper *orientation* of the raw quartz for cutting blanks. The manner in which a wafer is cut from the mother crystal determines nearly all of the operating characteristics of the oscillator. This has to do with the fact that a piezoelectric crystal's reaction to an applied electric field depends on the relationship between the direction of the field and the orientation of the *electrical axes* of the crystal. Quartz crystals, by their very nature, possess a great deal of internal structure and general symmetry. Crystallographers describe the structure of crystals in terms of a set of axes, related through the symmetry of the crystal. A common descriptive system is the XYZ system. In this system, the principle, or optical, axis of the crystal is designated the Z axis. This is the axis about which the crystal displays "three-fold symmetry" (meaning the crystal can be rotated about this axis into any one of three identical orientations, each separated by an angle of 120°); it is also the direction of the pointed apex of well-formed crystals. Any direction perpendicular to the Z axis, bisecting the corner formed by two adjacent faces of the crystal, is defined to be an X axis. The remaining Y axis is perpendicular to both the Z and the X axes, emerging perpendicularly from a crystal face. Figure 4.1 illustrates these definitions.[40] The X axis is known as the electrical axis and the Y axis is known as the mechanical axis; an electric field applied along the X axis will induce a physical strain along the Y axis.

Determining the axis directions for a particular crystal is important as many of the crystal's physical, thermal, optical, and electrical properties depend upon which direction through the crystal you are considering. For instance, the speed of light within a quartz crystal varies depending on which direction the light is traveling. Crystals of this type are said to be "birefringent" and form double images of objects viewed through them. As mentioned above, the strength of the piezoelectric effect, and the type of physical deformation of the crystal, depends strongly on the orientation of the imposed electric field and the crystal axes. Some orientations give no piezoelectric effect, others result in an elongation of the crystal, while still others result in a shearing displacement of the crystal faces.[41] Furthermore, particular crystal oscillators are designed to have very precise thermal properties. As such, the

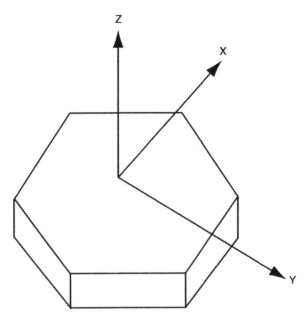

Figure 4.1. Definition of the XYZ crystal axis system

proper orientation of the final quartz wafer with respect to the original crystal's axes is very important.

Not all quartz crystals share the same internal structure; defined in terms of their optical properties, some crystals are said to be "right handed" and others "left handed." The "handedness" of the crystal refers to how the polarization state of light (the orientation of a light beam's electrical and magnetic fields) changes or rotates as it passes through the crystal along the Z axis. The handedness of a quartz crystal is important in the manufacture of certain types of oscillators.[42] To make matters even more complicated, a single quartz crystal can form in such a way that certain portions are right handed and others are left handed. This is known as "optical twinning." Much more important to oscillator production is the similar state known as "electrical twinning." In electrically twinned quartz, different regions possess opposite electrical characteristics. Obviously, such electrically twinned material is completely unsuited for use in a crystal oscillator. During initial inspection of raw quartz, the inspector checks both for optical and electrical twinning. Whereas optical twinning can be detected in most crystals by the use of polarized light, electrical twinning cannot. To search for electrical twinning, a slice must be made through the raw crystal and then etched in strong acid. The appearance of the etched surface will differ depending on the electrical orientation of the different regions (see Figure 4.2).[43]

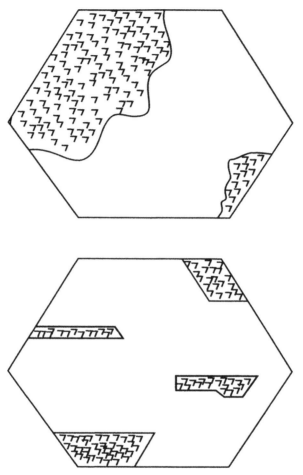

Figure 4.2. The appearance of twinning in etched quartz slices: electrical (*top*) and optical (*bottom*)

Determining the direction of the three axes can be quite simple if the particular crystal is a textbook example possessing well-defined faces and pointed ends. In such cases, the axes are simply defined according to Figure 4.1, above. However, natural crystals are rarely textbook examples. Many, due to the conditions under which they grow, never form smooth faces and pointed ends. Others have these surfaces eroded and broken off over the eons. One particular type, known as river (or unfaced) quartz, possesses the completely smooth surface common to stones found in bodies of running water. For crystals such as these, other techniques must be utilized. One such mechanical means is known as orientation by parting. In this method, a slice taken perpendicular to the Z axis is heated and then rapidly cooled in a beaker of water. Fractures then appear in the slice, which are oriented parallel to the X axis. Optical methods, such as polarization measurements are successful in some instances. Also, etching and electrical polarity measurements are sometimes successful.[44] However, the most accurate method of orientation utilizes X-ray measurements.[45]

Beams of X-rays passed through or reflected from the surface of a wafer of quartz will display patterns characteristic of the internal atomic structure of the crystal. This process, known as X-ray diffraction, allows for highly accurate and precise measurements of the crystal structure. The X-ray beams are only reflected or transmitted strongly when they intersect the surface of the quartz at very exact angles with respect to the atomic planes. Thus, through either photographic or electronic means, measurements of the angle between the X-ray beam and the crystal planes can be carried out. By rotating the sample of quartz until a maximum amount of X-rays is detected, the locations of the crystal axes are determined. Along with being extremely precise, X-ray measurements are also quite rapid (with WWII-era measurements taking 10 to 15 seconds on average), greatly facilitating mass production of crystal oscillators.[46] Once the correct orientation for the particular quartz oscillator to be manufactured is determined, the mother crystal can then be sawed into slabs. The slabs can then be tested for electrical twinning; those that pass are ready for the remaining processing steps.

Most of the methods of determining the proper orientation described above were quite tedious and time consuming and, as many trial cuttings might be required, resulted in a large fraction of the mother crystal being wasted. Though the most reliable and efficient method for quartz orientation made use of X-ray measurements, only the Western Electric company was using X-ray orientation to any degree at the start of the war. A primary impediment to the use of X-ray equipment by the industry at large was cost; a fully equipped machine (including instruments for measuring angles) could cost nearly $3,000 and take months to acquire from General Electric, the primary manufacturer.[47]

In August 1942, Faber and Knights wrote to the QCS requesting assistance in acquiring an X-ray machine for their company. As the QCS had only a very basic knowledge of their company's operation, Major Olsen replied asking for a full description of their plant with regards to numbers of workers, the types of processing steps they carried out, and how many contracts they held and with which prime contractors. The issue of their request for an X-ray machine would be settled based on their response to these questions.[48] Faber decided not to wait for help from the government. He owned a very old portable medical X-ray machine, and having served in the medical corps during the First World War, knew how to operate it. With this crude machine, he worked out a method of photographing the crystal structure of his quartz blanks (using Knights' gas station restroom as his laboratory and darkroom). Using a very well-oriented blank as a guide, Faber would cut a wafer from a mother crystal, X-ray it, and compare the X-ray image with that taken from the properly oriented blank. By rotating the test X-ray to align with the guide, he could determine with very good precision how the mother crystal needed to be reoriented on the saw.[49] Faber kept his operation a secret until his reputation for producing extremely high-quality crystal blanks brought representatives from Galvin to his shop demanding to know how he did it. Faber ultimately collaborated with General Electric on improvements in the design of their X-ray equipment.[50]

The Signal Corps discovered very early on that simply pouring money into the established crystal plants for more workspace, equipment, and employees would not alleviate the crystal oscillator procurement crisis. New techniques and pieces of equipment needed to be developed to bring the production of crystal units from "laboratory scale" to true mass production. Attempts were made to bring about these changes from the "top down," with QCS men working with the large crystal producers and equipment manufacturers. However, what made the industry a true success was the amount of "bottom up" innovation that occurred. A great deal of the needed techniques and pieces of equipment were developed in the small shops scattered throughout the country. The primarily small, flexible companies and the men who ran them, for the most part technically savvy, hands-on, do-it-yourself independents, made the crystal industry ripe for this kind of evolutionary growth.

The efforts at advertising and more fully exploiting these innovations carried out by prime contractors such as Galvin and by the QCS and the Signal Corps labs made it possible for a single member's discovery to result in a substantial increase in production for the entire industry. Though probably the most important step in increasing the industry's production, ferreting out and publicizing innovative techniques or machinery was not the only occupation of the Quartz Crystal Section. The officers and technical staff were kept busy with inspection and troubleshooting visits, policing the industry for violations of production protocols or government regulations, attempting to protect valuable production men from the Selective Service, and striving to conserve the precious supplies of raw quartz currently being procured at great cost and effort. Though growing quickly, the crystal industry during the early years of the war still required a lot of support.

5

RIDING THE "FLAT WHEEL LIMITED"—OVERSEEING A MASS PRODUCTION INDUSTRY

One of the major questions plaguing the Signal Corps in early 1942 was whether General Roger Colton, a man with electrical engineering degrees from M.I.T. and Yale and long seen as strictly a research man, could overcome the immense supply problems facing him as chief of the Materiel Branch. With Chief Signal Officer Dawson Olmstead's 1941 reorganization of the OCSigO, the areas of Research and Development and of Supply were joined together under the Materiel Division. Though seemingly disparate functions, Signal Corps leaders had long viewed R&D and Supply as the beginning and end of a continuous process. Former CSigO Mauborgne had always believed that a piece of equipment should be completely developed and tested before it ever reached the production stage, and Olmstead firmly held that "Supply begins with Research and Development."[1] Mauborgne's slow pace of development, however, was one of the leading reasons for his early replacement and, with the coming of war, his successor faced even more challenging obstacles of supply.

In volume II of its official history, the Signal Corps is described as being, in January 1942, "not yet ready, not yet equipped with enough of anything."[2] This was certainly the case with crystal oscillators. The industry, as it stood during that second month of war, had a production capacity (as estimated by the previous year's output) of 100,000 units; barely 1% of the 9 million units needed for 1942.[3] In order to get this industry moving toward dramatically increasing its production capabilities, education would be the first order of

Crystal Clear: The Struggle for Reliable Communications Technology in World War II, by Richard J. Thompson, Jr.
Copyright © 2007 by Institute of Electrical and Electronics Engineers

business for the Signal Corps. A meeting was held on March 20, 1942, in Washington to find out just what the industry needed to know. Signal Corps officers, laboratory engineers, and representatives of the manufacturing outfits came together to air their concerns related to crystal production.

One of the immediate needs of the industry representatives was clarification on specifications. There existed a great deal of confusion with respect to the Signal Corps' specifications for the DC-9 and FT-243 crystal units. It was decided that further meetings would need to be held in New York and at Ft. Monmouth to discuss these issues. Other questions dealt with proper testing procedures and with methods of working with smaller pieces of raw quartz.[4]

Based on the results of this meeting, a course was designed for educating the industry in the currently held best practices for crystal unit production. The start of the course was delayed for several months as facilities were prepared at the rapidly expanding Ft. Monmouth. Finally, in September 1942, even though the facilities were not quite finished, invitations went out to all members of the industry as well as the Signal Corps depots and labs to send a representative to attend the first session of the production course.[5] The course consisted of 12-day sessions with classroom work and hands-on training in the machine shop. Students of the inaugural class were called upon to serve as instructors for the six subsequent sessions that were held.[6] Though the members of the Signal Corps Labs (SCL) continued to deal with a steady stream of requests for assistance throughout the war, the courses were discontinued once John Ziegler began offering his classes in Kansas City.[7]

"THE ACTION OF ONE PRODUCER IN WITHHOLDING FROM THE INDUSTRY INFORMATION PERTINENT TO THE BEST PROCESSING METHODS . . . MUST NOT BE TOLERATED"

Perhaps no other industry embodied Olmstead's belief in supply beginning with research and development than the crystal industry. Not only did the products themselves require extensive research and development work, but their manufacturing likewise demanded a great deal of effort to improve methods and machinery. As discussed earlier, these manufacturing research and development efforts were distributed throughout the industry itself, thus taking a portion of the burden off of the OCSigO and the Signal Corps Labs. Nevertheless, in the early months of the war, many of the companies that had developed new pieces of machinery or methods of production attempted to keep them private, treating them as "trade secrets."

In one particularly frustrating case for the OCSigO, Bliley Electronics even attempted to keep a new production step a secret from the Signal Corps itself. Over the course of the two weeks following the Pearl Harbor attack, Willard Clark and John Fill, radio engineers from the SCL, toured 11 crystal producers from New England to the Midwest. Overall, they found the production capacities of these plants to be quite low. It seemed to them that most of the operations visited could benefit greatly from the knowledge and experience of the more established crystal producers. In their report of this inspection tour, Clark and Fill made what is perhaps the first suggestion of an information pool. Thus, in this frame of mind, they were particularly amazed and outraged when they discovered that the Bliley Electronics Company was preventing Signal Corps inspectors from observing the finishing and calibration steps of their production line. Restricted to one particular room designated for their use, the Signal Corps inspectors did not get to see the quartz crystal units until they had been sealed in their holders.

As a primary subcontractor to the Galvin Corporation, Bliley was handling orders for tens of thousands of crystals, and doing so fairly efficiently. The company gave a great deal of the credit for their production output to their new methods of finish-grinding and calibration. However, they considered these methods to be "trade secrets" and would not permit anyone, even the resident Signal Corps inspectors, into the rooms where these steps took place. Even Clark and Fill were forbidden to view these rooms (at the decision of Dawson Bliley himself). The inspectors were simply told that the new grinding and calibrating processes were greatly simplified; so much so that relatively unskilled workers could be quickly trained to carry them out.[8] In fact, even this small amount of information was designed to mislead the inspectors. The process was not a grinding procedure at all; it consisted of using a strong acid to etch away the last layers of quartz needed to bring the blank to the proper oscillating frequency. It had been decided after its development by John Wolfskill, chief crystal engineer at Bliley, that the process would be called the "X-Lap Process" to throw any competitors who might hear of it off the track.[9]

Clark and Fill were assured by the Bliley managers that these restrictions on inspector access had been approved by the Signal Corps contracting officer in Chicago. This claim was investigated by the Signal Corps men as soon as they arrived at the offices of the Chicago Signal Corps Procurement District. No documentary evidence could be found to substantiate Bliley's claim, nor could the contracting officer remember anything about such an agreement. Furthermore, the officer had recently visited the Galvin plant and found several instances where Bliley-supplied crystal units had to be rejected. Among the problems found were cracked wafers and foreign matter within the holder, all problems that could have been caught by the inspectors at the Bliley plant had they been allowed to examine the units before they were sealed in their holders.[10]

In a strongly worded report to O'Connell, Clark and Fill condemned the actions of the Bliley management. In their opinion, "the action of one producer in withholding from the industry information pertinent to the best processing methods" resulted in a decrease in the total production of the industry. "This must not be tolerated," they urged. O'Connell agreed completely with this conclusion and asked the Production Expediting Section of the OCSigO to see what could be done. What could be done was get the Signal

Corps lawyers and those of Bliley together in the same room and have them reach a mutually agreeable conclusion.[11] As an intermediate step, Bliley did agree to give tours to Signal Corps officers, provided they signed a secrecy oath first. Ultimately, in the words of Charles Bliley (author of the Bliley Company's official history), Bliley gave in to "political and patriotic pressures" and agreed to share their etching-to-frequency process with anyone who was interested.[12]

As it turns out, though etching was eventually incorporated into the Signal Corps requirements, it was not until the summer of 1944 that this process became a standard processing step (and only then in response to the "aging crisis"; see Chapter 10). Most of the small crystal plants found that converting to etching was too expensive during the early years of the war. As a postscript to this episode, O'Connell attempted to sooth the ruffled feathers of the Bliley Company and repair any damage done to their relationship with the Signal Corps. After their agreement to release the details of their production methods, O'Connell sent word to John Fill that, as he had won, he could "afford to be magnanimous and tell them he didn't mean to say anything to give them any offense." Fill had "offended the Blileyites," and now O'Connell wanted only "Peace—Sweetness—and Light."[13]

The Bliley "X-Lap" episode was an extreme and isolated case. For the most part, crystal producers were quite willing to share their knowledge and expertise with other members of the industry. Sometimes the Quartz Crystal Section actively sought out their assistance.[14] At other times, plant owners contacted Washington directly, offering to share information on a new processing step or testing procedure.[15] As networks of subcontractors began to take shape, a great deal of information sharing took place within them. Ernest Ruff of Kansas City recalled many instances where he was loaned to other plants to help them with their production problems and, on other occasions, to obtain from them ways to overcome his own company's difficulties.[16]

"IT IS RECOMMENDED THAT THIS MACHINE BE INVESTIGATED..."

Improving the current sawing technology was one of the first objectives of the pre-war industry. Producers such as A.E. Miller, Louis Patla, Theodore Valpey, and G.C. Hunt were all involved in developing homemade saws to speed up their production lines.[17] Grinding the blanks once they were sawed from the mother crystal was perhaps the most time-consuming production step and represented the most critical need for improvement if the making of crystal oscillators was ever going to become a mass production enterprise. If crystals were going to be produced in quantity, they would have to be ground in quantity. Perhaps the first true innovation in crystal production was G.C. Hunt's development (and P.R. Hoffman's later improvement) of the planetary lap, a machine for grinding several blanks simultaneously.[18] In this machine, the blanks moved in small circles as the grinding plate revolved about a central axis. The blanks, secured in holders of either Bakelite or metal, moved between two stationary lap plates. By employing multiple laps with different abrasives, blanks could simply be transferred from one lap to another as they went through the grinding stages. A plant utilizing five or six planetary laps in this manner could be expected to produce over 6,000 blanks per day, with less than a 10% rejection rate.[19] Hunt's work on grinding machines continued into the war years. Shortly after Pearl Harbor, Hunt made available to the Signal Corps his latest development, a machine to automatically grind up to 20 crystals to within a few kHz of their final frequency. Furthermore, the frequencies of the crystals could be tested while they were being ground. Hunt was more than willing to share his design with the rest of the industry, "provided his interests [were] adequately protected."[20]

Another crystal producer with a great deal of experience with cutting hard materials was E.A. Blakeley of Indianapolis. A former manufacturer of tombstones, Blakely entered the crystal industry shortly after the beginning of the war, specializing in the cutting and rough grinding of quartz blanks. A June 1942 visit to his shop by Major Olsen of the QCS found him developing a much improved saw (claimed to be able to cut so precisely that only one side of the blank would need to be ground) and a machine to rough grind 100 blanks at a time. Olsen was very impressed with Blakeley's work and recommended continued interaction between him and the QCS. Blakeley also expressed a willingness to share his designs with the rest of the industry, provided that the Signal Corps reimbursed him for his developmental expenses (around $1,000). Olsen recommended to his superiors, however, that they wait on this request until the machines had been completed and put into service.[21] From the summer of 1942 on, most of the Quartz Crystal Section's development work on grinding and lapping machines was carried out by Dick Stoiber. He spent a great deal of time on the road visiting crystal producers and commercial outfits alike, testing, evaluating, and reporting on the value of the various new models and their potential aid to the crystal industry.[22]

Along with supporting the equipment development activities of the individual crystal producers, the QCS played a supporting and expediting role for many commercial machinery firms. A great deal of assistance, for instance, was given to the Felker Manufacturing Company. As the nation's largest manufacturer of diamond-tipped saws, Felker was a very important supplier to the crystal industry. As such, it was important that they were kept operating at peak efficiency. At times, the QCS found it necessary to intervene on Felker's behalf to obtain raw materials (such as copper) needed for their production lines. At other times, large orders for Felker saws were placed by the SCL at Ft. Monmouth in order to assist the company in gaining higher, priority ratings (thus improving their ability to procure scare raw materials).[23] Other companies that produced saws, grinding machines, or other pieces of equipment applicable to the crystal industry, such as the Birdsboro Machine Company of Reading, Pennsylvania, the William A. Hardy & Sons Company of Fitchburg, Massachusetts, and the American Instrument Company of Silver Spring, Maryland, also received similar assistance from the QCS.[24]

Probably the most important among the "other equipment" category was that of X-ray machinery for the orientation of raw quartz. As discussed earlier, the use of X-ray images to correctly orient the mother crystal for the cutting of blanks was perhaps the most important technological innovation to the manufacture of quartz oscillators. Not only did X-ray orientation save a great deal of time, it also decreased tremendously the amount of quartz wasted during the cutting process (more on the conservation of raw quartz will be discussed below). At the beginning of the war, Western Electric was the only company making large-scale use of X-ray orientation. This is primarily due to its being the only company large enough to afford such technology. A primary goal of the QCS was to get X-ray machines into every crystal facility involved in cutting raw quartz. This meant that the machines needed to be made less expensive and easier to use.

Originally, General Electric was the only company to produce X-ray machines in any quantity. A second party, the Philips Metalix Company, also began producing the machinery, much faster and slightly cheaper than General Electric.[25] In fact, it was Philips Metalix that was asked to contribute photos of X-ray machinery for publication in the *Handbook for the Manufacture of Quartz Oscillator Plates*.[26] The Bendix Radio Company developed a set of accessories for use with the basic X-ray machines that proved very helpful in measuring the angles of the crystal axes of the raw quartz. The QCS worked out an agreement in which these accessories were manufactured by both Philips Metalix and General Electric, becoming standard equipment on their X-ray machines.[27]

Work on improving the accuracy and precision of X-ray technology continued throughout the first year of the war, being carried out by the manufacturing companies along with scientists and engineers at the Naval Research Laboratory and the SCL. In late October of that first year, a proposal was made by the QCS to fund a study by the Bartol Research Foundation of Swarthmore, Pennsylvania. Working in close collaboration with Philips Metalix, Bartol would endeavor to develop an extremely sensitive and accurate X-ray detector.[28] Offers of expertise also came in from academia. One example was J.W. Hickman, an instructor of chemistry at the Johns Hopkins University. Having several years of experience in working with X-rays, Hickman wrote to O'Connell in April 1942 offering the use of his X-ray laboratory for any studies the QCS might need to have carried out. Along a similar line, the QCS also contacted members of the scientific community with the hopes of stimulating research projects related to quartz crystal units, occasionally holding conferences with small groups of researchers.[29] Research was not limited to X-ray technology, nor even to equipment in general. As early as April 1942, the American Jewels Corporation of Attleboro, Massachusetts, was awarded nearly $55,000 to carry out a study to develop "standardized and uniform production methods" of crystal oscillator manufacture. The study would cover all aspects of the process, from initial inspection of the raw quartz to the final lapping and testing of the oscillator blank.[30]

All of the innovations, commercial developments, and scientific research in the world would do the industry no good unless the results could be incorporated into the standard production practices of all of the crystal producers. The *Technical Bulletins* and the *Crystal Round Table* publications were helpful, but of primary need to the early manufacturers was face-to-face interactions with Signal Corps personnel: trained technicians, engineers, and scientists from the Quartz Crystal Section and the SCL who could instruct, troubleshoot, and generally hold their hands as they attempted to get their plants up to maximum production.

"I SPENT MANY NIGHTS ON TRAINS BETWEEN WASHINGTON, CHICAGO, AND KANSAS CITY"

In the spring of 1943, Virgil Bottom was teaching physics at Colorado A&M University. Due to the extremely low population of male students on campus, his teaching load consisted primarily of a course he called "Physics for Home Ec." One day, during a laboratory session, he was called out of class by the department secretary, who informed him that he had a telephone call from Washington. Reaching the office (after a rather anxiety-filled walk down the corridor), Bottom found that the caller was Dr. Karl Van Dyke. The former student of Walter Cady and quartz researcher was now Chief Physicist of the QCS.[31] Constantly attempting to improve the "under-manned state of the Quartz Crystal Section," [32] Van Dyke had been working the academic grapevine looking for prospective scientists and engineers to recruit. One day during a conversation with Dr. Ralph Sawyer of the University of Michigan, Bottom's name was mentioned as someone in whom Van Dyke might be interested. Bottom had received his master's degree in physics under Sawyer's direction and had made a very good impression on the professor with both his research and his teaching abilities.[33] Now, as he interviewed Bottom on the phone, Van Dyke inquired as to his knowledge of crystal oscillators. Bottom described his experiences working with Herb Hollister and assured Van Dyke that he was quite familiar with both their theoretical and practical aspects. Van Dyke's response to this information was short and simple: "How soon can you come to Washington?"[34]

Though the offer was intriguing, Bottom had to worry about, among other things, supporting his wife and three young sons. Though Van Dyke assured him he would do everything he could to arrange for an acceptable salary, he warned Bottom in a follow-up letter that a person's salary didn't go very far in wartime Washington. The high cost of living coupled with the shortage of housing might make the transition difficult for him and his family. "On the other hand," he wrote, "there are very real problems to be solved and it is a challenge to the individual to meet them." Along with his follow-up letter to

Bottom, Van Dyke wrote on the same day to Dr. Lewis Webber, Bottom's department chair at Colorado A&M. Webber had written to Van Dyke in the hopes of keeping Bottom in Colorado. Professors, particularly of physics, were in short supply and Webber definitely did not want to lose Bottom. Van Dyke sympathized with Webber, but with Bottom's experience with crystal oscillators and his teaching skills (which would be utilized fully within the crystal industry), Van Dyke and the QCS could not help but cast "eager eyes" toward him. After making up his mind to join the QCS, it took Bottom nearly a month to complete all the necessary Civil Service procedures.[35] Delaying his start until after the end of the spring semester, Bottom boarded a train in Fort Collins, Colorado, on the morning of May 7, 1943. He reported for duty at the newly completed Pentagon two days later.[36]

Dick Stoiber described his first weeks of work with the QCS in 1942 as "wild and wooly." Bottom found the situation at the Pentagon a year later little different. A truly colossal building complex, it nevertheless became filled to capacity very quickly.[37] The bustling environment, the rampant inefficiency, and the mysterious logic of military procedures were somewhat overwhelming to a man who had spent most of his adult life in academia.[38] Bottom wouldn't spend that much time in the Pentagon; he spent a much larger fraction of his first year of service on the road interacting with members of the crystal industry.

The men (and some women) of the QCS spent a great deal of their time traveling. Their work was of such a priority that they were issued blank booklets of travel orders, allowing them a great deal of flexibility in responding to the needs of the industry. What they were not given, however, was unlimited funds for room and board. A QCS staff member having to cover three meals and a hotel room could easily spend more than their $6 per diem allotment (being "rewarded with a negative income" in the words of Virgil Bottom). If, however, the night were spent on a train between cities, one might actually spend less than the per diem ("positive income"). Trains were by far the more common mode of transportation, a mode that many of the QCS men found less than comfortable. Of particular dread was having to travel on the Pennsylvania Railroad, nicknamed the "Flat Wheel Limited."[39] In one particularly ironic episode, Bottom and a Signal Corps officer needed to get quickly from Wright Field, in Dayton, Ohio to Chicago. Offered seats on one of the air base's planes, they were given parachutes, five minutes of instruction in their use, and hurriedly took their seats. As the crew went through their preflight instrument check, however, it was discovered that the plane had a nonfunctioning radio. The Signal Corps men were forced to take the Flat Wheel Limited overnight to Chicago.[40]

The services offered by the QCS were divided between two units: a Technical Orientation Unit and an Industrial Engineering Unit. The Technical Orientation Unit offered assistance with machines and methods. They could recommend to producers the proper types of machinery needed for crystal manufacture. They could also assist in the training necessary to operate these machines and in the proper methods of utilizing them in a production scheme. The Industrial Engineering Unit offered advice on plant design and layout, along with management procedures. Of primary importance during the early months of the war was increasing the production capacity of each and every crystal producer. Whether by assisting with plant expansions, finding skilled plant managers, or providing direct technical and engineering aid, the QCS had to find a way of meeting the production goal set for the end of 1942 of 200,000 crystal units *per week.*[41]

Facility expansion was the first and most obvious means of increasing output. Basement operations, while fine for the amateur crystal trade, would not suffice for supplying the Armed Forces. Companies were strongly urged by the QCS to expand their facilities "to the limit" in order to increase production.[42] Members of the QCS helped directly with problems related to such expansions (including locating available facilities, working with the RFC and other agencies to secure funding, even assisting with legal issues such as having structures condemned so that crystal plants could remove them and build on their sites).[43]

Along with expanded facilities came the needs for larger work forces and for qualified plant managers to oversee their work. Whereas labor was available in the newly tapped and quite capable female work force, experienced plant managers were not. On occasions, members of the QCS Industrial Engineering Unit served as temporary plant managers until permanent replacements could be found. Sometimes such replacements were discovered working at other, larger, companies that could afford to share the men (or lose them entirely). In such situations, the QCS worked with the companies involved to facilitate the sharing or outright transfer of such managers.[44]

Companies were exhorted to utilize their expanded facilities around the clock. In August 1942, Olsen wrote to the crystal producers urging such utilization. In a letter that opened "This country is at war!" Olsen urged all manufacturers to go to a full schedule of three eight-hour shifts per day.[45] The majority of the quartz producers, both large and small outfits, eventually reached this utilization goal. The introduction of weekly production reports in June 1942 allowed the QCS to monitor the industry's consumption of raw quartz, its production of finished oscillators, and the radio manufacturers to whom these oscillators were shipped. These data enabled them to detect problems and dispatch technical or engineering advisors before they became too serious.[46]

Problems did occur that the QCS field men were required to troubleshoot. On some occasions, the advice of the QCS men seemed to fall on deaf ears. One example is that of A.E. Miller. Though his was one of the very first commercial crystal firms, Miller never seemed to make the change from the "lab scale" production methods, which served his amateur crystal business well, to the mass production techniques needed for wartime production. Though there were never any real complaints regarding the quality of his product, his company's constant shortfall in quantity caused the QCS a great deal of concern throughout the war.[47] Visiting QCS men would advise Miller of the recommended methods and equipment, but he seldom followed their suggestions. In fact, he insisted for some time in utilizing homemade equipment when much better models were available on the commercial market. In this particular case, at least, playing the role of the do-it-yourself independent operator did not yield dividends.[48]

It was the job of resident Signal Corps inspectors distributed throughout the industry to keep an eye out for problems in quality. Quite often, QCS men such as Bottom had to travel to plants in order to settle disagreements between plant managers and Signal Corps inspectors. Sometimes the problems arose from differing interpretations of the unit specifications. The QCS men would have to attempt to diffuse the situation and get the plant back into operation. On other occasions, real problems in production did occur.[49] Production would be halted and no further units from the offending company would be accepted until the problems were worked out. Sometimes, QCS men were temporarily assigned to the companies to directly oversee the corrections to the production problems.[50] In some instances, entire orders were canceled due to a company's inability to overcome problems and get into production.[51]

When threatened with losing its Signal Corps contracts, it always helped for a company to have friends in high places. The Federal Telephone and Radio (FT&R) Corporation, of Newark, New Jersey, was a company that never managed to reach production goals (in spite of a great deal of assistance from the QCS and a $164,000 plant expansion financed through the Defense Plant Corporation (DPC)). Even with 643 employees and running two shifts a day, the company had a backlog of unfinished crystals as of August 1943 worth over four million dollars.[52] The situation became so bad by the end of 1943 that the War Production Board (WPB) threatened to cut off their supply of raw quartz. Upon being notified of this, the QCS dispatched Virgil Bottom to the plant to investigate. Before leaving, however, he was stopped by Roger Colton, who told him that he was "confident" that Bottom "would find nothing wrong."[53] Colton was an associate of a Mr. Sosthenes Behn, the president of International Telephone and Telegraph Corporation (for which FT&R was a crystal-producing subsidiary). At times during the previous year, Colton had telephoned Behn personally to inquire as to the company's production problems and to encourage him to do all he could to improve the situation.[54]

At the FT&R plant, Bottom found a hopelessly disorganized production line. A great deal of raw quartz had been sawed, but never ground into blanks. Making matters much worse, even though X-ray equipment was available, most of the slabs seemed to have been sawed with the wrong orientation, making them useless for oscillator production. Bottom's report on the plant pointed out two additional shortcomings: excess breakage of quartz blanks, and a general lack of technical knowledge on the part of the production force and the apparent underutilization of the technical expertise of their engineer (a Mr. Hawk).[55]

Bottom recommended that the company be denied any more raw quartz shipments until it could demonstrate that it could overcome its current problems.[56] In his report to T.M. Douglas, the manager of FT&R, Major Swinnerton made no such recommendation. He simply pointed out the Signal Corps' concern over the inefficient use of raw quartz and stressed that, in the future, the "facility be used as effectively as possible."[57] However, the amount of crystal orders placed with FT&R for the following year *was* cut by 40% and recommendations were made by the QCS for the excess equipment purchased by FT&R through the DPC to be offered for sale to other manufacturers.[58] Years later, Bottom still believed that FT&R should have been shut down and that it was only through the direct intervention of Roger Colton that it was not. Though this charge might possibly be true, it's also likely that Colton felt it important to maintain any and all plants that were producing crystal oscillators, regardless of their productivity.[59]

This was a very important concern. The quartz industry had been built up throughout 1942 at a great deal of effort and monetary expense. To the QCS, the loss of manufacturers, even those of obviously lesser ability, could scarcely be afforded. This point, however, was not as clear to other areas of the Signal Corps. In fact, by spring of 1943, the crystal industry was actually believed by some to have "over expanded." An April 1943 report on the crystal situation by the Resources and Production Division of the OCSigO described the industry as such and claimed that a great deal of "salesmanship" would be required for crystal requisitions from the using arms to reach the 20 million mark estimated by the Signal Corps for the year's needs.[60]

The QCS disagreed with this view of an excess production capacity. Captain E.W. Johnson, Officer in Charge of the QCS in April 1943 believed that more orders for crystal units would be forthcoming and that all manufacturing facilities should be kept in operation. Johnson believed it made sense to keep the entire industry running, even if below full capacity, so that production could be easily increased in response to later emergencies.[61] Through the judicious distribution of Signal Corps contracts, along with similar efforts on behalf of primary contractors such as Galvin, the industry was kept going through the lean months of the spring and early summer of 1943 without significant disruptions in the crystal plants' operations.[62] This approach was indeed the correct one to take, as the number of crystals actually produced in 1943 exceeded 20 million and continued to increase throughout the remaining two years of the war.

A similar situation was faced by the QCS during the fall of 1944, though this time it was the manufacturers themselves who seemed concerned about an excess of industrial capacity. The successful invasion of France by the Allies in June 1944 signaled for many the imminent conclusion of the war. Radio and crystal manufacturers alike seemed hesitant to take on any more large orders, fearing that the war would end before they could finish production and be compensated. The QCS worked hard to dispel these ideas and keep production at peak capacity.[63]

"THE CRYSTAL DEPARTMENT HAS BEEN THE STEPCHILD . . . "

So, the technical advisors of the QCS continued to travel the country. Whether investigating possible additions to the industry, or troubleshooting problems for the current members, the QCS men visited nearly every section of the country. Sometimes the QCS responded to direct requests for assistance from the manufacturers, sometimes to warnings from the WPB, and sometimes they discovered problems themselves during inspection visits. One particular case of the latter involved the crystal operation of the Bendix Radio Corporation. The Baltimore division of the corporation had "mushroomed" during the first year of the war from 600 employees housed in a single building to 5,600 workers and several buildings. A portion of the Baltimore division was devoted to the production of quartz oscillators. An August 1942 visit to the crystal facility by Lieutenant Miller and Samuel Gordon found the operation less than they had expected for such a large and successful company. It appeared to them that the crystal operation was definitely the "stepchild" of the corporation, being staffed by men "found incapable on other production lines."[64]

The production of the plant was quite low, with a great deal of waste taking place. It seemed to Miller and Gordon that the primary reason for this was the lack of a trained crystallographer to oversee the orientation and cutting operations. Furthermore, it appeared to the QCS men that no one even knew just how much raw quartz was on hand and in what grades. Miller and Gordon instituted a precise inventory control system and recommended a particular Bendix employee to be designated "Chief of Orientation." They believed that this person had the skill and energy to get the production problems under control.[65]

By October, it appeared that a dramatic change, at least in attitude, had taken place at the Bendix plant. In the two months since Miller's and Gordon's visit, it seems that Bendix had reached the conclusion that they were not only a successful crystal operation now, but one of the best in the industry. Wallace Richmond of the QCS visited the plant to find out why they were canceling their orders for Felker diamond saws. Richmond was informed by the plant manager that these saws (considered state of the art by most members of the crystal industry) were inadequate for his use. He had a number of suggestions for modifications and informed Richmond that "the Felker saws could not be used on his production line unless they were rebuilt to his specifications." This demand was backed up with the claim that Bendix, being "the only producer of quartz crystals who is production minded" had managed to take "all the mystery out of the production of quartz crystal."[66] Confronted with this attitude, there was really nothing that Richmond could do.

By the first of December, visits to the Bendix plant seemed to suggest that things had settled down and crystal production was progressing (though problems did exist with the manufacture of particular types of crystal units).[67] On December 7, Carl Bertsch of the QCS visited the plant to check up on two rush orders for the Air Forces—one due to be shipped the next day and the other one week later. In consultation with the Bendix Planning Department, the final details of the first rush order were worked out and instructions for completing the order were telephoned to the crystal plant. The next evening, Bertsch visited the plant and discovered that the order had not been completed. It appeared that the telephone instructions had been misunderstood. To make matters worse, no one of authority was currently present at the plant. Taking matters into his own hands, Bertsch set about giving instructions and getting the production line up and running. The workers carried out their tasks quite efficiently and the deadline was missed by only a few hours.[68]

This was not the first time that Bertsch had had to assume "command" of the crystal operation at the Bendix plant. In a meeting with members of Bendix management the following morning, he informed them that he did not enjoy doing this, but that he felt that the alternative of missing a deadline for an Air Forces rush order was much less desirable. The Bendix team expressed their gratitude for Bertsch's actions the night before and explained that management problems related to the rapid expansion of the Baltimore facility were to blame for these instances. A reshuffling of management responsibilities was carried out and promised to prevent any future problems of this sort. One week later, Bertsch was back at the Bendix plant having to troubleshoot production problems related to the second Army Air Forces rush order.[69]

In his follow-up report to his QCS superiors, Bertsch expressed disbelief at the situation he encountered upon arriving at the plant, especially "after the splendid outlook of the week before." This time, the problem was due to a lack of blanks for grinding to the final frequencies needed by the Air Forces. The plant had essentially come to a halt and the Bendix managers did not know how best to proceed. A plan to delay shipment slightly while new sources of quartz blanks was obtained was agreed to by the Army Air Forces procurement office. Permission was also gained for only partial shipment of another rush order needed by the Ground Forces. It seems that not quite all of the "mystery" had been removed from the manufacture of quartz crystal oscillators by the Bendix Crystal Division. Fortunately for all involved, the QCS had been able to catch these various production problems and affect solutions before things got too far out of hand.

On other occasions, it seems that the inspection efforts of the QCS were not as greatly appreciated by plant owners as was the case with the Bendix corporation. One illustrative situation took place at the Frequency Measuring Service in Kansas City, Missouri. It seems that a Signal Corps representative visited the plant while the production engineer, Joseph Egle, was away. The plant was experiencing some production problems, and Egle was visiting nearby plants in search of solutions. The Signal Corps inspector was not pleased with what he found at the plant and word soon got around among the employees that the plant might very well be shut down. When Egle returned, he found his plant in "crisis." In a letter to the QCS he explained the source of the production problems and what had been done to address them. He further complained about the actions of the inspector. Overall, Egle felt that the situation could have been handled much better by the Signal Corps representative. He felt that inspectors should not "act like hunters or spies," and most certainly should not divulge their findings and possible actions to the employees. In his opinion, this all simply led to the production of rumors, not quality products.[70]

One last example demonstrates how the QCS and the crystal manufacturers could work together to solve efficiently critical problems related to crystal production. The British were developing a new radio transmitter that was slated to be used in "an important operation under tropical conditions" during the first quarter of 1945. The equipment had been in production for some time, but, in October 1944, a problem related to the radio's crystal units arose. When first designed, it appeared that a fairly standard crystal unit would be utilized. However, later changes to design and operating specifications raised concern that the crystal unit initially considered would not suffice. Tests would need to be run to see whether the oscillator could be modified to meet the new requirements. As the Signal Corps Laboratories at Ft. Monmouth lacked the facilities to conduct the test, a manufacturer was desired to carry out the tests in collaboration with Signal Corps engineers. Arrangements needed to be made quickly as the British were very eager to have production of the units commence (so eager, in fact, that they were willing to pay cash for the units in order to avoid the delays of lend-lease).[71]

The Henry Manufacturing Company of Los Angeles was approached to carry out the production tests. As his company currently produced the unmodified units, R.E. Henry agreed to take on the job. Virgil Bottom was selected to oversee the program. He and a representative from the British Ministry of Supply Mission in Washington would travel by airplane to Los Angeles and carry out the needed tests. Before leaving, Bottom requested that accommodations be reserved for himself and Lieutenant Colonel Coles in Los Angeles. Upon their arrival, Bottom and Coles were driven to the Bel Aire Hotel where they were led to their rooms along a path that wound through a grove of citrus trees and over a small brook. Used to much sparser accommodations, Bottom was somewhat amazed at the extent of his room (which included "two double beds and a bathroom larger than most motel rooms"). Also found in his room was a card stating that the room's rate was $16 per day. As Bottom's per diem totaled only $6 and Cole's only $4, the two men shared one room that night and relocated to the Hollywood Hotel for the remainder of their stay (rooms there only cost "$3.00 per day with no extra charge for the mice that invaded the waste baskets at night").[72] Though his plant was currently experiencing labor problems (a strike by workers had recently prevented him from even entering his own facility),[73] Henry managed to carry out the tests needed by the Signal Corps. The results suggested that the needs of the British government could be met with this particular type of crystal unit and plans for the manufacture of the needed quantity were immediately put into effect.[74]

"IF YOU FELLAS CAN DEMONSTRATE THE ABILITY TO MAKE CRYSTALS, WE WILL GIVE BOTH OF YOU DEFERMENTS"

One of the most serious threats to maintaining the production capacity of the quartz crystal industry came from the U.S. government itself: the Selective Service draft. Throughout the war, to the extent that it could, the QCS fought a battle to protect factory owners, plant managers, and skilled male employees from being drafted into the Armed Forces. Requests for draft deferments began coming in to the OCSigO almost as soon as the Selective Service legislation went into effect. The granting of deferments, however, was not within the powers of the Signal Corps; the ability to grant deferments lay primarily with the local draft boards. The Signal Corps could request deferments, however, and did on numerous occasions on behalf of crystal producers.

Though most were old enough to be exempt from the draft, some oscillator manufacturers were younger than the upper limit of 36 years. This was the case for Leo Meyerson and Al Shideler. They looked into the possibility of draft deferments before they ever agreed to start their crystal business. During their initial contacts with the Signal Corps, they made it plain that they "didn't want to invest a lot of money getting set up if there was a possibility of being called up for service." The procurement officer in Philadelphia to whom they were speaking assured them: "We need crystals badly. If you fellas can demonstrate the ability to make crystals, we will give both of you deferments."[75]

Another man in nearly the same situation was Stanley Valinet of Indianapolis. In the process of starting a crystal plant as a subcontractor to A.E. Miller, Valinet also wanted assurances that his efforts would not be rendered futile by the draft. Though already rejected once due to problems with his feet and eyes, he knew the requirements could always be lowered later in the war to the point where he would qualify. During a visit to his plant in June 1942, Harry Olsen, newly appointed head of the QCS, accompanied Valinet to the office of the local draft board, where he attempted to explain the importance of men such as he to the war effort.[76] This seems not to have satisfied Valinet, as he wrote to Olsen three months later still concerned about his draft status. In his reply, Olsen attempted to reassure him of the importance of his work. Many crystal workers ("especially key personnel") had already been granted draft deferments by their local boards. If he were still concerned, Olsen suggested he show this letter to his draft board to further substantiate the importance of his business to the war effort.[77] The efforts of Olsen, if not Valinet's foot and eye problems, were sufficient to keep him out of the draft; his Tru-Lite Research Laboratories continued to produce crystals for the Signal Corps throughout the duration of the war.[78]

Primarily, it was experienced crystal workers (particularly supervisors) that the QCS was urged to protect from the draft. One of the plant owners most active in requesting assistance with deferment problems was A.E. Miller. The files of the OCSigO contain numerous letters from him complaining of the loss (or potential loss) of valuable employees.[79] Miller was by no means the only one complaining; losing skilled male employees to the draft represented a potential business crisis for most of the manufacturers. The QCS wrote numerous letters to local draft boards on behalf of crystal workers. These letters (and a few telegrams in the name of the Chief Signal Officer) pointed out the crucial nature of the products manufactured by these men. A "critical occupation," crystal production was of the "utmost importance" to the Signal Corps. Furthermore, these men could be "of much greater service to the war effort by remaining" with their companies than they could be as members of the Armed Forces.[80]

Companies were instructed on the proper procedures for securing deferments for their employees. The first step was to retain an attorney to prepare affidavits attesting to the amount of war-related work being done by the company and the role played by the employee.[81] Furthermore, the time and expense necessary to train a replacement (if available) should be mentioned.[82] By July 1942, the QCS was suggesting that these employees be described to their draft boards as being impossible to replace (which, in fact, they were).[83] For those companies whose employees were still selected by the local boards for induction, an appeals system did exist. The QCS distributed to the entire industry a letter outlining the proper procedure to follow for appealing a local draft board's ruling. The first step was to send a letter to the local board requesting a hearing before the Appeal Board. The Appeal Board should be sent copies of all previously submitted documentation. Should the Appeal Board rule against them, an employer could then petition the State Director of Selective Service. As a great deal of requests to the QCS for assistance involved men that had only received induction notices, employers were urged to contact the OCSigO only after following all of these steps.[84] Even before the need for such efforts became necessary, the employer should file an application with the U.S. Employment Service requesting workers for their plants. This application would serve two purposes. First, it just might result in their gaining qualified employees to replace any lost to the draft. Second, if no such qualified personnel were available, this would strengthen any appeal they might later have to make.[85]

In order to help spread the word about these procedures, large contractors, such as Galvin, were asked to make sure that all of their subcontractors were aware of the danger of losing their employees and of the necessary steps to be taken to protect them. The QCS kept these prime contractors informed of any changes to the process that took place as the war progressed.[86]

During the summer of 1942, the QCS wrote up a detailed list of the types of workers employed in crystal production. This list was then sent to the War Man-Power Commission for their consideration and (hopefully) inclusion in their Dictionary of Occupations considered essential to the war effort.[87] Olsen attempted to educate the members of the Man-Power Commission as to the importance of the crystal industry, describing crystals as one of "the most critical of all items being produced to win this war."[88] The list was also forwarded to the crystal producers so that they could use the proper job descriptions when stating the type of work done by their employees.[89] The list consisted of the following 15 job types:

Plant Owner
Plant Manager
Plant Foreman
Assistant Plant Foreman
Plant Supervisor
Plant Engineer
Production Manager
Plant Inspector (i.e., inspector of raw quartz)
Quartz Crystal Orientor
Quartz Crystal Cutter
X-ray Technician
Quartz Crystal Edger
Quartz Crystal Rough Finisher
Quartz Crystal Final Finisher
Tests and Inspectors (of blanks and finished products)[90]

The education program of the QCS with regards to the Selective Service appears to have been successful. From late 1942 until late summer of 1943, no further draft-related correspondence appears in the QCS files. In August 1943, however, one last letter went out from the QCS to the crystal industry informing them of the increase in the number of men being drafted and its possible consequences for the crystal industry. The letter, written by Lieutenant Colonel Messer, advised the manufacturers of the existence of a new network of Signal Corps Labor Offices. The purpose of the Labor Offices was to "aid, advise, and inform all producers and manufacturers . . . on problems arising from the National Selective Service Act and the induction of personnel into the Armed Forces." Included in the letter was a list of the addresses of the Labor Offices and their territories of responsibility.[91] One last possibility presented by the QCS for crystal workers being drafted (the joining of crystal grinding teams) will be discussed in Chapter 9.

Crystal plant owners and employees were not the only men whose deferments were important to the QCS; it found it necessary to petition for the deferment of many of its own scientists and engineers. When Virgil Bottom received his draft notice from the Wichita Draft Board a few weeks after joining the QCS, he was informed that he was already AWOL. He spent a few more nervous weeks while Colonel O'Connell worked on getting his deferment approved.[92] (As a side note, Bottom had attempted to join the Navy a year earlier in hopes of getting an officer's commission. The recruiter just couldn't seem to understand what a physicist was and what use one would be to the Navy. Ultimately, he informed Bottom that "the Navy is mighty particular who it takes in." Bottom then replied "and I am mighty particular what I join," and walked out of the recruiting office.) In its deferment petitions for its civilian members, the QCS (often over the signature of the CSigO) pointed out that, along with being of great use to the crystal industry and being essentially irreplaceable, these civilians could represent the Signal Corps "to officers and high ranking technicians of other army agencies and to the management and senior engineers of many contractors" much more easily than could a member of the Service.[93] Furthermore, being civilians, they could be stationed at a single location for essentially an indefinite period of time (an option not available to Army officers or enlisted men).[94]

When one considers the range of equipment that utilized crystal control, and the difficulty in procuring the quantities of crystal units needed, it is easy to see the importance to the Signal Corps of gaining draft deferments for these crystal workers. In the past, some authors have appeared to disagree with the deferment process. In his book *Band of Brothers*, Steven Ambrose discusses Eisenhower's desire to go back on the offensive after the battle of Bastogne. The primary problem, however, was a lack of sufficient manpower. "The United States had not raised enough infantry divisions to fight a two-front war. This was a consequence of the pre-war decision by the government to be lavish with deferments for industrial and farm labor, and to refrain from drafting fathers."[95] To a foot soldier in Europe during the winter of 1944–1945, the deferment program might well have seemed "lavish." Nevertheless, if all of the crystal workers receiving deferments had been drafted instead, they probably wouldn't have numbered enough to constitute a single additional company. Furthermore, the performance and dependability of the radios carried on the backs of the platoon radio men and in the armored vehicles and tanks (due in large part to the manufacturing skills of these deferred men) had a much greater impact on the progress of the war than the addition of another rifle company ever could have.

Not all crystal manufacturing employees received deferments. One in particular was George F. Fisher of the Valpey Crystal Company. Joining the company as a schoolboy in 1940, Fisher worked part time at the crystal plant until he left home for college. After finishing his first year of college, he returned home to work full time at Valpey and await the "imminent draft call." He was drafted in 1943 and sent to Replacement Radio Operator School at Camp Crowder, Missouri. This occupation did not impress Fisher with its possibilities for "longevity of life or good health." Aware of the Signal Corps' involvement with the crystal industry, Fisher called Wallace Richmond of the QCS (an acquaintance from his time at Valpey) and asked him to pull some strings for him. After only a couple of weeks, Fisher received orders to report to Ft. Monmouth. With his personal effects being shipped ahead, Fisher was ordered to stop off at the Holabird Signal Depot in Baltimore and deliver a package to the commanding officer. Thinking he would simply deliver the package and continue his trip to New Jersey, Fisher was surprised to find himself immediately reassigned as an instructor in the crystal grinding school recently activated at the depot (it took some time for his personal belongings to get reassigned as well).

The Holabird Depot was responsible for training the new crystal grinding teams being sent oversees in response to the aging crisis (more about this in Chapter 9). After training a few teams in the crystal grinding arts, Fisher found himself assigned as the chief noncommissioned officer of a unit headed for the China-Burma-India Theater of Operations. After 18 months overseas, Fisher returned to the United States for Officers Candidate School. Shortly thereafter, the war ended and he returned to the crystal business, making it his life-long career.[96] As demonstrated by George Fisher's story, sometimes, even those crystal men who were not able to gain deferment from the draft still managed to serve in a crystal-related capacity.

"I DON'T NEED ANY MORE EQUIPMENT FOR TESTING"

Just as important as producing a sufficient quantity of crystal oscillators was ensuring the quality of the units. Testing accompanied the entire manufacturing process. The raw quartz was tested for defects that might negatively impact any oscillators made from it. Quartz blanks were tested for any temperature sensitivity on the part of their frequency or activity. Lastly, the finished oscillators were tested under actual operating conditions. All manufacturers were expected to carry out the proper tests applicable to their particular products. Prime contractors also retested samples from subcontractors and the Signal Corps Inspection Agency was ultimately responsible for certifying finished units.

Temperature testing was very important, as a large portion of the operating specifications for oscillators dealt with the temperature range over which the crystals must function. Furthermore, the temperature-related characteristics of a crystal wafer were related to the particular type of cut made from the mother crystal. Problems with orientation thus could be discovered through temperature testing. As the temperature ranges in the specifications were fairly large, special equipment needed to be used; both dry ice freezers and large ovens were used to cover the specified range. Tests were supposed to be run on the units every few degrees throughout the entire range. Though a tedious and time-consuming process, the Signal Corps placed a great deal of importance on this procedure. A manufacturer who did not follow these procedures could find his entire production line shut down until he was able to carry out the tests properly.

One isolated exception to this, however, took place at the Daughtee Manufacturing Company in Chicago. A crystal-finishing operation contracted to the Galvin Corporation, Daughtee had always supplied a quality product. Nevertheless, a Signal Corps inspector newly assigned to the plant attempted to shut them down due to their completely unorthodox method of temperature testing. Lew Daughtee employed no precision testing equipment; he didn't even have a thermometer. What he did do was take each day's production and place them in an ice box and freeze them overnight. The next day, a worker would take them from the freezer and test their frequency response and activity. The units would then be placed in an oven and baked. While still too hot to handle with bare hands, the units were tested again. Those that passed both tests were shipped to Galvin. When informed that he would have to install the proper test equipment and start following the correct procedure, Daughtee replied that he didn't "need any more equipment for testing." His opinion was, since his units worked at temperatures much colder "and a hell of a lot hotter" than the lower and upper extremes of the specifications, they would surely work just as well within that specified range.[97]

The Signal Corps needed proof. Extensive tests were carried out at the Galvin plant of recently received Daughtee units. Out of perhaps 2,000 units, borderline failures were observed in only three. It was very hard to argue with results such as these. The response of the Signal Corps inspectors was not to "push the point about temperature," but to find out "how the hell he does it." Daughtee was only too happy to demonstrate his finish-grinding operation. Nothing there seemed any different than observed at any other crystal plants. "How about the cutting operation?" the inspectors wanted to know. For that, Daughtee explained, they would have to visit the man he bought all of his blanks from, a Mr. Leon Faber. It was this investigation of the Daughtee company that led to the discovery of Faber's homemade X-ray orientation process.[98]

Along with temperature testing, finished units needed to be tested under actual operating conditions. Initially, this meant having on hand at the manufacturing plant a working model of the transmitter or receiver in which the unit would be used. With the growing number of crystal manufacturers and the rapidly expanding inventory of Signal Corps radio equipment, this method of final testing became impossible to support. What was needed was equipment to mimic the operating conditions of the various radios. Such test sets could be distributed among the manufacturing industry and used in place of the actual sets.

The development of such test sets took place primarily at the SCL. The Signal Corps employee most involved in the development of the sets was Marvin Bernstein. A ham radio operator since 1933, Bernstein was hired to work at the Ft. Monmouth laboratories two weeks after the Pearl Harbor attack. His initial work focused on finding acceptable methods of frequency control that did not involve quartz crystals. By 1943, however, industry production was meeting demand, so this line of research was abandoned and Bernstein was transferred to the crystal branch. Under the direction of Captain C.J. Miller, and in collaboration with C.B. "Lanky" Davis and A. Pritchard, Bernstein became involved in the test set development program.[99]

The primary goal of the test sets was to mimic the electrical characteristics of the actual radios. If the overall capacitance, inductance, and resistance of the radio could be reproduced by a simple set of electronic components, the operation of the crystal unit under these conditions could be observed.[100] Furthermore, a much greater degree of testing uniformity could be achieved across the industry.[101] Early versions of a "universal test set" were developed at Bell Labs but did not really satisfy the needs of the industry. The first successful models were developed at the SCL. Two of the first sets used, the CES-1 and the CR-8/u, employed adjustments in capacitance or resistance to model the particular radio of interest. As the vacuum tubes contained within the test sets could change their operating characteristics over time, it was necessary for Signal Corps personnel from the Quality Assurance Section to travel the country with standard calibration sets, making sure that the test sets at the manufacturing plants were operating properly. Ultimately, Bernstein and his group hit upon the idea of controlling the test set's characteristics through its overall impedance. The crystal impedance (CI) meter ultimately answered the need for a uniform test set, flexible enough to operate throughout the entire frequency range currently being utilized by Signal Corps radio equipment.[102] The CI meter became the standard piece of test equipment for crystal units and was used by the industry for the next two decades.[103]

"IN VIEW OF THE EXTREMELY CRITICAL RAW QUARTZ SUPPLY SITUATION . . ."

As will be discussed in Chapters 6 and 7, the primary responsibility for supplying the crystal industry with raw quartz fell to agencies of the civilian government, such as the War Production Board and the Foreign Economic Administration. However, the conservation and proper utilization of that raw quartz by the industry came under the responsibility of the Signal Corps. Throughout the war, the QCS was involved in a constant effort to educate manufacturers in the needs for and the methods of conserving raw quartz. One of the most successful methods for conserving raw quartz during the manufacturing process was the X-ray orientation procedure discussed above. Along with correct orientation, efficient sawing was next in terms of importance. The thinner and sharper the saw blade, the less quartz that was wasted. A letter was distributed by the QCS to the entire industry in December 1942 urging "maximum efficiency" in orienting and sawing techniques. The letter ended with a warning that companies "not efficiently utilizing present stocks" of raw quartz would find themselves in "extreme disfavor" when the time came to more strictly allocate it.[104] A follow-up letter calling the industry's attention to the "serious import" of the previous letter was sent out a month later. The letter called for a "very careful review" of each manufacturer's orienting and sawing practices. It also included a chart describing the best practices in this area with which the companies could compare their own methods.[105]

A third major improvement in crystal technology that played perhaps as important a role in quartz conservation as the machinery mentioned above was the development of crystal oscillators utilizing smaller quartz wafers. Much of the early research regarding the use of smaller oscillators was carried out at the Bliley Electric Company under the supervision of John Wolfskill. During the summer of 1937, Wolfskill experimented with constructing oscillators using a quartz wafer approximately half the size of that currently used in their amateur crystal units. It was found that, in terms of activity, the smaller crystals were actually better than the larger ones for certain frequency ranges. At all other frequencies, the performance of the new crystals was equal to that of the larger ones. One possible drawback to the smaller crystals was that the capacitance of the electrodes and leads would now account for a much larger fraction of the overall capacitance of the unit. This would require much greater attention to the design of these components.

In 1940, with quartz now being considered a "strategic material," Wolfskill reviewed this research in a Bliley internal report. While the smaller crystals used less raw quartz, their use would require a "complete redesign of [the] entire line of mountings and holders" for the crystal units. Also, the handling, grinding, and final lapping of the smaller units would require "more care." However, in Wolfskill's opinion, the savings in raw quartz might "more than compensate" for the increased production difficulties.[106] The use of smaller crystals was immediately seen as a godsend by Procurement Planning officers such as Zermuehlen. If crystal units could be made with smaller crystal wafers, the supply of raw quartz available to the industry could, for all intents and purposes, almost double instantly. The development of oscillators utilizing smaller crystals was initiated during the first year of the war for both Ground and Air Forces radio sets.[107]

Along with utilizing less quartz, and minimizing waste, the QCS attempted to conserve quartz by keeping it entirely out of the hands of unqualified manufacturers. By late 1942, Signal Corps labs were instructed to strongly discourage new applicants from entering the crystal business. It was felt by the QCS that the potential waste that would result as they learned the business would outweigh the potential benefit of their products.[108] Sometimes these parties attempted to appeal this policy, even asking their senators and congressmen to intervene on their behalf. This rarely appears to have been successful, however.[109] On other occasions, aspiring plant owners who possessed technical skills were urged by the QCS to join already established crystal outfits.[110]

Along with attempting to keep unqualified manufacturers out of the crystal business, the QCS was also involved in policing the industry. Along with ferreting out and ridding the industry of sub-standard manufacturers, and attempting to get sluggish producers up to speed, the QCS was also on the lookout for possible violations of contracting policies and ordering priority regulations.[111]

The hard work of the Signal Corps of building and supporting a mass production industry for crystal oscillators during 1942 and 1943 paid off handsomely. By January 1943, the 125-company industry was consuming in a month more than twice the amount of raw quartz imported in all of 1939. Production that same month was 1.6 million units, over 22 times that of February 1942, with a commercial value of $125 million. At the beginning of 1944, the industry's production had reached the level of two million units per month. Improved techniques and concepts were being developed so fast that a new idea could become obsolete within a matter of weeks. Furthermore, the number of oscillator wafers that could be produced from a pound of raw quartz in 1943 was almost double that of 1940. All of these dramatic improvements came at a cost to the Signal Corps of less than one million dollars.[112]

The human cost, however, was incalculable. Tremendous numbers of man-hours were spent in strategy meetings and planning conferences, in touring crystal plants, and riding the Flat Wheel Limited. This work was absolutely essential; without it, the Signal Corps would never have been able to supply the Armed Forces with the mobile, reliable communications equipment that it desperately needed. As alluded to earlier, the Signal Corps was not the only agency hard at work supporting the military's need for crystals. The civilian government played just as large a role. Without the supplies of raw quartz, whose mining, purchasing, and transportation to the United States were accomplished largely due to the efforts of the Miscellaneous Minerals Section of the War Production Board, the industry built by the Signal Corps would not have been able to produce a single oscillator.

6

SUPPLYING A MASS PRODUCTION INDUSTRY—THE CIVILIAN GOVERNMENT STEPS IN

During the late 1930s, as the using arms of the military cried out for crystal control, and the Signal Corps laboratories attempted to answer those cries, the members of the OCSigO staff involved with procurement continued to sound warnings regarding the potential supply of raw quartz suitable for radio oscillators. With no known domestic deposits, and the quite probable interruption of imports from Brazil during wartime, the continued acquisition of a sufficient supply of raw quartz was anything but certain. These fears were very well founded; the military was indeed taking quite a risk with its steadily increasing dependence on crystal control. Fortunately, elements within the civilian government would soon begin to tackle the immense problem of accumulating the raw materials that would be needed for war. As it turns out, these actions came just in the nick of time.

"TO PROVIDE FOR THE COMMON DEFENSE BY ACQUIRING STOCKS OF STRATEGIC AND CRITICAL MATERIALS..."

On June 7, 1939 (just three months before the German invasion of Poland), Congress passed the "Strategic and Critical Materials Act (Public, No. 117)." Concerned with "acquiring stocks of strategic and critical materials essential to the needs of industry" for the production of items necessary for "common defense," it addressed both the identification of these essential materials along

Crystal Clear: The Struggle for Reliable Communications Technology in World War II,
by Richard J. Thompson, Jr.
Copyright © 2007 by Institute of Electrical and Electronics Engineers

with their accumulation.[1] Ultimately, such materials came to be defined as "those required for essential uses in a war emergency, the procurement of which in adequate quantities, quality, and time is sufficiently uncertain for any reason to require prior provision for the supply thereof."[2] In the meantime, the naming of a particular item as a strategic or critical material would be done by the Army and Navy Munitions Board (ANMB) under the authority of the Secretaries of War, the Navy, and the Interior. In cases where the domestic supply of such materials was deemed insufficient, the Procurement Division of the Treasury Department was empowered to purchase the needed amounts from foreign sources. A sum of $100 million was appropriated to cover needed expenditures through June 30, 1943.[3]

One year after the enactment of Public 117, the next major step toward preparing the nation to defend itself was carried out by Congress with the passing of the "Act to Expedite the Strengthening of the National Defense" on July 2, 1940. Of particular importance to the development of stockpiles of needed raw materials, Section 6 of the Act authorized the President to "prohibit or curtail" the exportation of strategic or critical materials. For those items so named by a presidential proclamation, a license would have to be acquired before they could be exported from the country.[4] The first such proclamation was made the same day that the Act was passed. Along with the "arms, ammunition, and implements of war" defined in previous proclamations, lists of "basic materials," chemicals, aircraft parts, armor plating, and machine tools were given. Among the 26 "basic materials" listed was quartz crystals (both piezoelectric and optical).[5] Furthermore, when the ANMB released its official list of "Strategic or Critical Materials" three months later, quartz was one of the 32 materials listed.[6]

With this new emphasis on war preparedness, a multitude of governmental management and advisory agencies sprang into existence. One that became involved with the quartz procurement effort was the National Defense Advisory Commission (NDAC). Formed May 28, 1940, just weeks before the fall of France, the NDAC was tasked with assisting the massive billion-dollar defense buildup asked of Congress by President Roosevelt two weeks earlier.[7] Work related to the stockpiling of raw quartz, directed by Dr. R.J. Lund, began later that summer. On the first of August, C.R. Avery of the Western Electric Company was invited to Washington for a meeting with the NDAC and representatives of the Treasury Procurement Division and the Bureau of Mines. Before any real work could begin with regards to organizing and increasing the accumulation of raw quartz, an estimate had to be made of the amount needed. The purpose of this meeting was to determine the proper size of a quartz stockpile needed to cover two-year's worth of wartime demand. Once the size of the needed stockpile was determined, the issues of purchasing, importing, inspecting, and distributing the raw quartz would need to be addressed.

Working from estimates supplied by the Signal Corps, a figure of 102,000 lbs was agreed upon for the size of the two-year stockpile. Added to this amount would be the 4,900 lbs currently held in Navy stocks, bringing the size of the stockpile to 106,900 lbs. C.B. Hamilton, with the Treasury Department, informed the group that approximately $600,000 was still available for the purchase of raw quartz. As the industry and government stocks amounted to approximately half of the 106,900 lbs, this would be adequate to complete the purchasing, at current prices, required for building the stockpile.[8]

Shortly after this meeting, the Committee for the Procurement of Quartz Crystal of the NDAC was formed, with Avery being named as Chair. This committee, composed of representatives from the larger quartz manufacturers, the Treasury Department, Bureau of Mines, and the ANMB, would struggle with the details of procuring the large amount of quartz needed for the stockpile. The United States had never been involved in purchases of the amounts currently needed and the contracting procedures followed by the Treasury Department were now considered much too slow to efficiently build up the stockpile. Furthermore, it was the opinion of the committee that the only way to appreciably increase the amount of Brazilian quartz available to the United States would be to dissuade the Brazilian government from selling quartz to other countries. This would, in effect, serve two purposes. Along with increasing the amounts available to the United States, it would decrease the amounts available to potential enemies. At present, the largest importer of Brazilian quartz was Japan, and it was feared that much of this quartz was ultimately shipped to Germany through Russian ports. It was also agreed that sending a small party of government buyers and inspectors to Brazil would better insure the purchase of quality material at fair prices.[9]

Furthermore, the committee felt that the specifications governing the types of quartz that could be purchased were "somewhat more severe than necessary."[10] A meeting was called for the following week to discuss these specifications in detail. Along with two Signal Corps representatives, the Quartz Advisory Committee produced a set of specifications at that meeting that, while still guaranteeing "a very good grade of quartz," would permit the more efficient purchase of larger amounts of crystals. Larger crystals, and those having two or more natural faces were, of course, more desirable than smaller or unfaced crystals. However, smaller, unfaced crystals could make up approximately 15% of any lot purchased. Also, crystals containing obvious defects could be purchased provided the usable volume of the stone was at least 60%.[11]

Based on the results of these two meetings, an official Quartz Crystals Procurement Plan was submitted to the Chief Signal Officer for his approval on September 23, 1940. The plan outlined the current state of the stockpile, taking into account military and industrial holdings, along with the 14,800 lbs either already acquired or under contract by the Treasury Department under the auspices of Public 117. As of September 1, 1940, a deficit of approximately 40,000 lbs remained of the 106,900 lb proposed stockpile. Discounting the industrial reserves (which were in the hands of a very small number of companies and would not be available to the industry at large in time of crisis), an amount of 87,200 lbs of raw quartz was needed to be procured by the Treasury. To facilitate this, a small staff of experts was again proposed to be sent to Brazil to work closely with quartz exporters. Furthermore, the State Department would be requested to ask the Brazilian government to allow the purchase of quartz by foreign countries only after offering the minerals to the United States.[12]

As it stood, the plan was a good one. Given time and an accurate estimate of the amount of quartz needed, it would have succeeded. In reality, the government would have to acquire the needed quartz in the absence of both of the aforementioned requirements. Barely a year was left before the "emergency" referred to in the planning documents would engulf the country. Furthermore, those tasked with acquiring the raw quartz had to work with continuously increasing estimates of the amount needed.

"IN SUMMARY, YOUR PROPOSED STOCKPILE IS ONE WHICH IS IMPOSSIBLE TO MEET PRIOR TO ABOUT 1950"

Nothing could be done about the timing of the coming war. However, it seems reasonable that the amount of quartz needed could have been estimated. Nevertheless, throughout the remainder of 1940 and through most of 1941, such estimates were found to be nearly impossible to agree upon. The primary problem was the lack of knowledge regarding the numbers of crystal-controlled radios that would be needed by the using arms and the number of crystals each set would require. The decision to utilize crystal control in the series of radio sets being designed for the Armored Forces during the fall of 1940, along with the rapidly increasing needs of the Air Forces, quickly made the summer's stockpile estimates obsolete.

The Signal Corps, primarily the Procurement Planning Section under the direction of Lieutenant Colonel Bogman, continued to work toward an accurate estimate of the military's quartz crystal needs. Computations made in late September 1940 suggested the crystal needs of the Armored Forces would require over 36,000 lbs of raw quartz.[13] Much harder to estimate were the needs of the Air Forces. Compounding the problem was the difficulty the Signal Corps had in getting realistic estimates from the Air Corps as to the number of planes that would utilize the various crystal controlled sets. Furthermore, the possible number of crystals that would ultimately be required for each of these sets ranged from 2 to 50 (a difference of over 57,000 lbs in the amount of raw quartz needed for manufacturing the oscillators).[14] In the end, the Signal Corps chose to err on the side of caution, using the larger estimates of crystals per set in their calculations.

On December 10, the CSigO submitted to the Assistant Secretary of War the Army's estimated crystal needs for a two-year emergency (309,103 lbs). Compared with the estimates generated by the ANMB 13 months earlier, the new estimate represented a 300% increase in the amount of quartz needed. To say the least, this caused a fair amount of consternation on the part of the War Department. In his reply to the CSigO, Brigadier General H.K. Rutherford, Director of the War Department's Planning Branch, clearly and patiently explained to Mauborgne the problems with his request. First of all, at the current rate of the Treasury Department's procurement program, it would take 5 years to complete the originally estimated stockpile; 20 years for this revised plan. Furthermore, the amount of quartz now requested would require purchasing the entire output of Brazilian radio grade quartz for three and a half years. This assumed that the United States would be able to buy all of Brazil's production. This was not the case at present; Japan and other nations (including our potential ally, Great Britain) were still purchasing raw quartz from Brazil. General Rutherford closed his letter with a recommendation that the Signal Corps pare their requirements "down to a reasonable minimum" and begin work on purchasing stocks, possibly through the Reconstruction Finance Corporation.[15]

In response to this setback, Bogman directed the Procurement Planning Section to institute plans for securing between 40,000 and 100,000 lbs of quartz while more accurate estimates of actual needs were generated.[16] It is amusing to imagine the War Department's reaction upon receiving this revised estimate and finding that it included an *additional 150% increase* over the 1939 ANMB estimate. The two-year estimate submitted to the War Department in March of 1941 totaled over 450,000 lbs. This time, the response letter from Rutherford did not refuse the request outright, but demanded detailed information on how the estimate was calculated. Information was also requested regarding possible design changes that would facilitate the use of smaller crystal plates. The War Department was interested in the current state of research regarding crystal substitutes, along with the possibility of lowering the raw quartz specifications even more so as to utilize the lower quality quartz available in much larger quantities from Brazil. This information would be necessary if the War Department hoped to justify the request to the ANMB, the Treasury Department, and the Office of Production Management.[17]

By the end of 1940, the NDAC had reached the end of its effectiveness.[18] As such, it was replaced on January 7, 1941, by the Office of Production Management (OPM). Intended to "increase production for the national defense through mobilization of material resources and the industrial facilities of the nation," the OPM was charged with (among other things) determining the needs of the War and Navy departments and was expected to "take all lawful steps necessary to assure the provision of an adequate supply of raw materials."[19] The quartz-related activities of the NDAC were transferred, still under the direction of Lund, to the OPM.[20]

Rutherford's warning about the need to justify increased needs of quartz to the OPM was quite correct. Even before official requirements were presented, the OPM got wind of the increased estimates. In mid-April, George

Moffett of the OPM wrote to the ANMB outlining the current situation with regards to quartz purchasing. Although contracts had been let by the Treasury Department for the remainder of the 102,000-lb stockpile, only half of the amount had been received (and a great deal of that material had been rejected under the current specifications). Moffett felt that procuring the amounts rumored to be desired by the military "would require more time than we can afford to spend." Furthermore, Moffett pointed out, such an increased demand for crystal oscillators might be more than the current crystal industry could handle (this point is especially prescient considering Zermuehlen's discovery of a "deficit of production capacity" was over six months away). Decisions regarding the wide-scale utilization of quartz crystals would need to be made "promptly."[21]

Another point requiring prompt action by the ANMB and the Signal Corps was the issue of purchasing specifications. The majority of the quartz mined in Brazil was of a size and quality that prevented its purchase by the Treasury Department Procurement Division. This material was not refused by the Japanese and Italians (and to a lesser extent, the British), however. Active purchasing programs had been underway by these nations for some time. Though the export of many of its natural resources, including quartz, had recently come under stricter control by the Brazilian government, Moffett was not sure whether quartz exports to Japan would be halted. If so, this might actually work out to be detrimental to the United States as the loss of its largest customer for low-grade quartz might lead to the collapse of the entire Brazilian quartz mining industry. To better understand the entire issue of specifications, Moffett informed the ANMB, C.B. Hamilton of the Treasury Department was preparing a report for all interested agencies.[22]

Hamilton's report was submitted to the ANMB three days later. In it, Hamilton claimed the current specifications to be the chief impediment to the accumulation of adequate stocks of raw quartz. Contrary to popular belief, he had evidence that Japan was indeed using the low-grade quartz it purchased from Brazil in the manufacture of crystal oscillators. Hamilton blamed the large companies represented on the NDAC Quartz Crystal Committee with intentionally skewing the specifications to require the larger, easier-worked crystals utilized in their own plants. Only with the assistance of representatives from Bliley had Hamilton been able to include any reference to smaller, unfaced quartz in the original specifications. Accusing the larger companies of creating a "myth" regarding the nonusability of low-grade quartz, Hamilton pointed out the fact that these companies subcontracted with smaller companies such as Bliley, A.E. Miller, and Standard Piezo knowing full well that they utilized much smaller pieces of raw quartz. In his opinion, the specifications should be written to incorporate the types of quartz used by the smaller companies, the companies producing the majority of the industry's output. Hamilton suggested that the only true requirements for the acceptance of quartz for oscillator manufacture be that the stone be optically clear and weigh more than 7/8 of a pound. All other desired characteristics should be listed as "preferences." In terms of usable volume (i.e., that amount of a raw quartz crystal that could be used for producing oscillators), the current 60% requirement was good, but, in his opinion, stones possessing as little as 40% usable volume should still be purchased (though at a disccunted rate).[23]

The ANMB submitted its final estimate of stockpile needs to the OPM in late April 1941, estimating the total needs of the Army and Navy at a little over 300,000 lbs per year. Furthermore, the ANMB report pointed out that the Treasury Department had expended all of the funds appropriated under Public Law 117. General Hines of the ANMB suggested that perhaps the Metals Reserve Corporation (MRC, a subsidiary of the Reconstruction Finance Corporation) might be solicited to carry on the quartz purchasing program.[24] Though much more than the estimates of 1939, the OPM accepted the Board's request. Furthermore, it contacted Charles B. Henderson, President of the MRC, and passed along Hines' suggestion. Compared with the original estimate of 102,000 lbs, the OPM now requested that the MRC attempt to acquire 900,000 lbs: 600,000 lbs for a two-year stockpile and 300,000 lbs for the current needs of the crystal industry. In addition, the revisions to the purchasing specifications suggested by Hamilton would now become official. The MRC agreed to take on the job, provided that Hamilton himself would leave the Treasury Department and come run the program for them.[25]

One week later, on May 22, a meeting was held in R.J. Lund's office in order to inform the Signal Corps, National Bureau of Standards (NBS), and the Bureau of Mines of the latest developments. Lieutenant Colonel Tom Rives, one of two representatives of the Procurement Planning Section, declared that both the new stockpile estimates and the revised purchasing specifications were acceptable to both the Signal Corps and the Navy. Hamilton informed the group that credits were being established in Brazilian banks to be used for the purchase of raw quartz. The credits were related to a very large loan that was currently being negotiated between the two countries. The initial purchases by the MRC would most likely encompass entire warehouse holdings; the usable quartz would be picked out and the waste discarded. This would have two benefits: first, the average cost of the quartz would be significantly less (approximately one-fourth the current rate) than if only high-grade material were being purchased. Second, this would maintain the market for small quartz; the continued mining of which would concurrently maintain the supply of higher-grade material.[26]

The other Signal Corps representative, Lieutenant J.E. Gonseth, inquired of the industry's ability to utilize the lower-grade material. Hamilton assured him that it was usable and that it was simply a question of manufacturers adapting their production lines to make use of it. Dr. Frederick Bates of the NBS suggested that a study be carried out to determine just what types of quartz were usable for oscillator manufacture. He offered the use of his own facility, but pointed out that such a study would require funding to hire personnel and purchase equipment. Lund felt that this was a good idea and suggested it be followed up on as soon as possible.[27]

Essentially, all of the uncertainties relating to the quartz program could be grouped into two categories: raw quartz acquisition and inspection, and the production capacity of the industry. With respect to the inspection of raw quartz purchased by the MRC, Bates assured the group that his operation at the NBS could "handle any quantity" that the MRC could send them. Unfortunately, the inaccuracy of this boast would come back to haunt the crystal program throughout the early years of the war. In order to determine the potential production capacity of the industry, Lund suggested having the OPM send out letters to the manufacturers asking for the pertinent information. Gonseth replied that the OPM was pretty much an unknown entity to the crystal manufacturers; a survey carried out by the Signal Corps would have much more potential for success.[28] This decision led to the survey carried out by Captain Zermuehlen (discussed in Chapter 3).

Throughout the summer of 1941, the MRC worked to get its purchasing program up and running. Leonard J. Buck of New York City was hired to serve as chief purchasing agent. As such, he would work with the exporters in Rio de Janeiro to purchase the raw quartz and then handle negotiations for its sale to manufacturers. Buck's efforts would take place in parallel with those of commercial importers; feeling that the business cartels created by the independent importers were too important to lose, the government allowed them to continue their work of directly supplying the crystal industry with raw materials. In June, as part of a larger loan package, the RFC signed an agreement with the Brazilian government to purchase up to four million pounds of raw quartz per year for two years (as this represented essentially the entire quartz production of the nation, the problem of competition with other countries was eliminated).[29] It seemed that the country was well on its way to building the needed quartz stockpile.

"AN IMMEDIATE INCREASE IN THE INSPECTION RATE OF THE BUREAU OF STANDARDS IS IMPERATIVE"

After the morning of December 7, 1941, all talk of a quartz stockpile to see the country through a future emergency became moot. The nation was now at war and every piece of quartz mined in Brazil was needed immediately by the crystal oscillator industry. Throughout the fall, the MRC had managed to build a stockpile of approximately 800,000 lbs of quartz; nearly the entire amount agreed upon the previous summer to see the country through two years of war. However, on the day the war finally commenced, no one knew just how much of those 800,000 lbs were of sufficient quality for producing crystal resonators.[30] The inspection program of the NBS, promised to handle whatever could be thrown at it, had utterly failed to keep up with the pace of imports. The inspection program was perhaps the single most critical link in the chain stretching from the mines of Brazil to the cutting rooms of the industry. Its shortcomings came closer than any other crisis to crippling the entire raw quartz supply program.

When the Treasury Department began its procurement of quartz under the Strategic Materials Act, inspection was a two-step process. Initial inspection was carried out at the Washington Navy Yard. Material found to be acceptable for further inspection was forwarded to the National Bureau of Standards. Under the original specifications, final inspection involved the cutting of raw stones to assess the crystal's electrical properties. This procedure could take up to *three days per crystal.* By the beginning of 1940, it was quite apparent to the Treasury that changes had to be made (at the current rate of inspection, it would take approximately 1.5 years to inspect the 14,800 lbs of quartz currently in the Treasury stockpile). The proposed solution was to divide the inspection duties among industry contractors and government labs, but, ultimately, only the NBS was given the inspection duties. Of the three industrial concerns asked to bid on an inspection contract, Graybar Electric gave no response at all, RCA formally declined to bid on the contract, and Western Electric's bid was considered unacceptable. The Navy Gun Factory also declined to bid on the contract, leaving the NBS as the only entity to proffer an acceptable bid.[31]

In their proposal, the NBS offered prices based on two methods of inspection: the standard "cutting" method and a method based on optical inspection alone. The optical method was intriguing for several reasons. First, requiring no cutting of the crystals, it was much quicker. Second, it allowed for inspection of every crystal in a purchased lot (in the cutting method, only 4% of the stones in a given shipment were inspected). Thus, better determinations of quality could be made (and, whereas the standard method required acceptance or rejection of the entire lot based on the 4% inspected, with the optical method only those individual crystals failing to pass inspection would be returned to the supplier). Whereas the Treasury Procurement Division favored the optical method suggested by the NBS, it had to gain the approval of the ANMB before the official inspection specifications could be changed. Receiving the request for this change in the specifications on April 30, the ANMB forwarded the suggestion to the OCSigO for its consideration two days later. The following day, Lieutenant Colonel A.A. Farmer of the Procurement Planning Section responded for the CSigO stating the Signal Corps' acceptance of both the NBS as the inspection facility and the proposed method of inspection.[32]

Under the direction of Frederick Bates of the NBS Optics Division, inspection got off to a very slow start. By August 1940, approximately 8,000 lbs of the 14,800 lbs contracted for by the Treasury had arrived at the NBS, but only 10% had been inspected.[33] By November, the Treasury had to begin delaying shipment of acquired quartz until the NBS had inspected a sufficient amount of the crystals already on hand. The possibility of finding commercial operations to share the inspection duties was again considered but not acted upon.

Shortages in equipment, trained personnel, and facilities were given as explanations for the large backlog of uninspected quartz that existed by the end of 1940.[34] The facilities problem, at least, was addressed by year's end with the inspection program being "moved into larger and more appropriate quarters."[35]

Another impediment to the efficient inspection of the raw quartz stockpile was the frequently changing specifications. The specifications governing the acceptance or rejection of raw quartz continued to evolve through May 1943, going through at least three iterations during 1940 alone.[36] Quite often, changes in specifications regarding size, structure, or weight of crystals required the reinspection of previously rejected material. Though these changing specifications did delay somewhat the inspection program of the NBS, the fact that Bates was active in the efforts to write the specifications means that he shares some of the blame.

The inspection of raw quartz generally involved two determinations: an overall acceptance or rejection, and a grading as to quality. Of primary importance to the first determination was the size of the overall crystal. Initially, stones were required to weigh two or three pounds in order to be accepted for oscillator work. This requirement was lowered to one pound after the start of the war and eventually to around four ounces.[37] Of secondary importance was the "cleanliness" of the crystal, i.e., how easily light passed through it. Though crystals could contain some defects, those accepted for the manufacture of oscillators needed to be mostly clear.

The quality grade of the crystal was based on just how many defects were present, along with the overall shape (large-diameter stones were preferred) and number of faces possessed by the crystal. Internal defects included cracks as well as "phantoms" and "needles" due to the inclusions of foreign materials or air bubbles. The internal structure of the crystals was examined using what is known as an oil bath. When simply observing a quartz crystal under normal conditions, the internal reflection of light from the faces of the crystal usually prevents the examination of its internal structure. When submerged in a bath of oil possessing optical transmission properties similar to that of quartz, light passes completely through the crystal without undergoing internal reflections. Thus, any defects or imperfections within the crystal could be observed. The oil bath examination often included the use of a very intense carbon arc lamp.[38]

Along with examining the quartz for internal defects, the inspector also looked for evidence of optical twinning using polarized light. Initially, both internal defects and optical twinning were believed to have detrimental effects on oscillator plates cut from the raw crystal. However, as will be discussed in more detail below, it was eventually discovered that any detrimental effects of these properties were minimal. After the completion of its inspection, a raw quartz crystal was placed into one of several quality grades. The grades were usually distinguished by just how much of the crystal's volume was usable for oscillator manufacture. Along with the changes in acceptable size and weight, the definitions of these grades went through a great deal of change throughout the first years of the war (by early 1942, the lowest acceptable usable volume had dropped from the initial 60% figure to 30%).[39]

With the increased quartz purchasing during 1941 under the direction of the MRC, the stress upon the NBS inspection program continued to build. Though over two million pounds of raw quartz was imported during 1941 (nearly half purchased through the efforts of the MRC), the amount of inspected and graded quartz available in the stockpile was only 13,600 lbs.[40] The entire quartz inspection staff of the NBS at the end of 1941 consisted only of 10 men, utilizing four inspection tanks. The OPM and the MRC had been pushing Bates throughout the year to increase the size of his staff. His response was that the availability of competent men was very limited. The truth of this statement depended upon the definition of "competent." For Bates, it meant a man who had a college degree in physics. For the Signal Corps, however, it meant anyone, male or female, with a high school education.[41] Bates also claimed that the inspection program was being hampered by the lack of sufficient storage and work space at the NBS. In response to this complaint, the MRC funded the construction of a storage warehouse.[42] Slowly, the inspection rate of the NBS program did increase. By going to a two-shift work schedule to better utilize the available equipment, quartz was being inspected and graded at a rate of 8,000 lbs per week by March 1942.[43] Still, this was nowhere near what it needed to be to keep up with the current supply, much less the amounts that would be arriving now that the nation was at war. By mid-year 1942, even though a year's worth of raw quartz existed in the stockpile, it would take over two years to inspect it at the NBS's current rate. Even worse, orders from manufacturers for raw quartz were now having to be shorted.[44]

The MRC continued to pressure the NBS to increase its production, even offering to fund the construction of a new inspection facility. At a meeting between the WPB, the MRC, and the Board of Economic Warfare, the point was made very clear by the WPB representatives "that an immediate increase in the inspection rate of the Bureau of Standards [was] imperative." Finances would not be a problem, Bates was assured; "any request for money for the inspection laboratory would be granted immediately."[45] The Signal Corps was very happy to hear of this offer. They had been making their own attempts to speed up the quartz inspection process, to little success. By June 1940, the Signal Corps was training quartz inspectors of its own.[46] In July, an offer was made to Bates to lend him nine Signal Corps men to work at his facility.[47] Bates turned them down, however, claiming a lack of space and the lack of personnel to further train the Signal Corps men.[48]

Things did improve somewhat during the latter half of 1942. Work began in late summer on the new inspection facility, though construction was held up several times due to arguments between Bates and the contractors concerning building materials (wood vs. steel) and the type of heating system (gas vs. coal furnace) the building would have. Bates did drop his college physics requirement for inspectors and, with an increased staff of 63 inspectors, put his operation on a three shift per day schedule.[49] By August, approximately 25,000 lbs of quartz was being inspected each week. Labor problems still existed for Bates, however, as several of his inspectors were lost to the draft during the fall of 1942.[50]

Problems still existed for the quartz procurement program, as well. By October, the industrial demand for raw quartz had reached 75,000 lbs per month, whereas the amount of usable quartz being released by the NBS each month was only around 60,000 lbs.[51] The General Development Division of the OCSigO did what it could to assist the inspection program, continuing to train Signal Corps inspectors and working to gain the proper purchasing priorities for the Laboratory so that it could receive needed inspection equipment (such as arc lamps and inspection oil) quickly.[52] Ultimately, an inspection facility was put into operation at the Toms River Signal Laboratory, where over 1.6 million pounds of raw quartz was inspected and graded through the first nine months of 1943.[53]

The new NBS facility in Washington, DC, was completed and occupied in December of 1942. With a staff of 166 inspectors and technicians, the potential output of the program was now estimated at 150,000 pounds per week. It appeared as though the quartz inspection bottleneck would finally be broken.[54] Other problems continued to hamper the program, however, most due to Bates' propensity to micromanage all aspects of the program. Although he had administrative duties related to several other areas of NBS research, Bates declined to delegate any of his authority over the quartz program. Not even the shipment of inspected quartz to manufacturers could take place without his direct involvement. On more than one occasion, significant delays in shipping resulted in the temporary closing of crystal plants. Had the laboratory supervisor (who enjoyed the confidence of the MRC) been able to authorize shipment in these cases, no loss of plant production would have occurred. Ultimately, it took letters to the Secretary of Commerce and a direct meeting between representatives of the OPM and the director of the NBS for Bates to loosen his control somewhat. Though he did not relinquish any official control, the laboratory director was given increased authority and the program ran fairly efficiently throughout the remainder of the war.[55]

The amassing of raw materials is useless if these materials cannot be transferred quickly and efficiently to industry for use in production. The quartz procurement efforts of the U.S. government were significantly hampered by the failure of the NBS to get its inspection facility adequately staffed and operating in a reasonable amount of time. A period of two full years went by before the inspection rate of the program reached anywhere near acceptable levels. While it is true that the facilities and staff of the NBS were not adequate in 1940, it is also true that they had every opportunity to avail themselves of assistance from the MRC and other government agencies to build additional facilities and increase their staff. Bates' stubborn refusal to accept anyone but college-educated physicists as inspectors (even when faced with losing many of these men to the draft) and to reject those offered by the Signal Corps was inexcusable.[56] Furthermore, his delaying of the construction of the MRC-funded inspection laboratory and his administrative micromanaging only added to the overall lackluster performance of the inspection program. In an enterprise characterized by hard work, innovation, and dedication to duty, the performance of Frederick Bates and the NBS inspection program in supporting the quartz crystal industry was definitely lacking.

"THE METALS RESERVE COMPANY, AN AGENCY OF THE GOVERNMENT OF THE UNITED STATES, IS HEREBY DESIGNATED AS THE BUYER"

With the coming of war came one more change in the governmental oversight of the nation's industry. As had the NDAC, the OPM failed to live up to its stated purpose. Hampered by a lack of real authority and clear leadership, it never was able to get the complex functions of priorities, scheduling, and supply under control. On January 16, 1942, President Roosevelt created the next (and final) agency to be charged with this responsibility. The War Production Board (WPB), headed by a single chairman, Donald Nelson, was composed of the Secretaries of War and the Navy, the Federal Loan Administrator, the Director General of the OPM, the heads of the Office of Price Administration and the Board of Economic Warfare, and the President's Special Assistant for the defense aid program. The Chairman of the WPB was to be a kind of economic czar, exercising "general direction of the war procurement and production program." Nelson would be in charge of setting the "policies, plans, procedures, and methods" of the Federal agencies and departments involved in the defense program.[57]

As it had been under the NDAC and the OPM, the quartz crystal program remained under the leadership of R.J. Lund, with his being named the Chief of the Miscellaneous Minerals Branch of the WPB. A Quartz Crystal Section was organized under Lund's branch to directly oversee the purchasing efforts. This section was headed by Dr. James Bell until his transfer to the Board of Economic Warfare in the summer of 1943. Bell's assistant, Robert McCormick, then assumed leadership of the section and directed its activities throughout the remainder of the war.[58]

Bell took office during a very chaotic time. As of January 3, 1942, the quartz holdings of the MRC amounted to over 1.8 million pounds. However, half of this stockpile was considered too small for use in radio oscillator production. Of the remainder, as discussed above, only a small fraction had been inspected. Furthermore, as the Treasury stockpile had never been completed, the question arose as to whether stocks from the MRC should be transferred to the Treasury to complete it. Though the MRC specifications allowed the purchase of crystals containing as low as 40% usable volume, the Treasury's requirements were for 60% usable crystals. In light of the fact that completing the Treasury stockpile from MRC holdings would essentially remove all of the highest grade material from the MRC stockpile, it was decided at a February 9 meeting between the WPB, Treasury, War, and Navy Departments that no such transfer would be carried out. The Treasury stockpile would be considered complete and all future government purchases would go to the MRC.[59]

By March 1942, the supply picture began to clear a little. Earlier estimates had predicted a 74,000-lb shortfall for the year. However, a more detailed survey by the Signal Corps found that, between the current supplies of the government, private importers, and industry, the needs of 1942 would be met. Thus, the MRC would focus its efforts on building a stockpile for the demands of 1943.[60] Brazil, in light of the previous year's loan agreement, had become a supermarket of raw materials for America's industry. Along with quartz, government agents and private importers began to contract for large amounts of minerals such as bauxite, calcite, diamonds, iron, mica, and tungsten. Food and other agricultural products also made their way to the United States in great quantities.[61]

As the chief purchasing agent for the MRC, all contracts for quartz purchases went through the office of Leonard Buck. The standard contracts were detailed, four-page affairs. Detailed instructions were given on packing and shipment along with an entire page related to specifications for acceptable crystals. Eighty percent of the contract price would be paid at time of delivery, with the remaining 20% being paid upon the satisfactory results of inspection at the Bureau of Standards.[62] This clause was particularly unpopular with Brazilian exporters due to the extremely slow pace of NBS inspections. Furthermore, as the standards of the private importers were somewhat more relaxed than those of the government, the exporters felt that they were very much at the mercy of the government inspectors.[63] With the establishment of inspection facilities in Rio in late 1943, this policy of delayed payment was ended (see Chapter 7).

"THE U.S. COMMISSION HERE ARE RUINING THE MARKET AND MAKING DAMN FOOLS OF THEMSELVES..."

Initially, the Brazilian exporters may have felt that the stricter specifications and delayed inspection required of sales to the United States worked against them. However, in the spirit of "buyer beware," it appears to have actually worked very well to their advantage. As early as September 1941, complaints regarding the effects of government contractors on the Brazilian quartz market were making their way to the Signal Corps. With the greatly increased purchasing efforts of 1942, things only got worse.

Initially, the complaints came from private importers. Though they competed among themselves, the dozen or so quartz importers shared a common goal of business success. Their new competition, the U.S. government, was not viewed as sharing that goal. Claims of purchasing worthless crystals for high prices filtered back to the American offices of the importing companies. One of the largest importers, Donald Murray, was infuriated by the actions of the government purchasing agents and attempted to spread the word. Writing to his brother in New York, Murray urged him to let the U.S. crystal manufacturers know what was going on. He hoped that their collective pressure could force the government to clean up the situation in Rio. Unaware of the MRC's policy of buying low-grade material in order to keep the mining operations in high gear (as well as to insure friendly political relations with Brazil), Murray believed the buyers were simply "making damn fools of themselves." Furthermore, it became harder for him to purchase quality material when the exporters could sell everything to the MRC buyers for much higher prices. Whereas the MRC felt it necessary to pay higher prices in order to acquire the amount of raw quartz needed for its stockpile, Murray saw the payments as evidence of "shady" deals, "kickbacks," and widespread corruption.[64]

By February 1942, the NBS was reporting cases of significant differences between the quality of the quartz described on invoice documents and the actual contents of the shipments. Often, boxes labeled as higher grade material contained very poor or even worthless stones.[65] Complaints continued to come into the OCSigO from importers, complaining of the greatly increased prices and lowered quality of crystals offered them by Brazilian exporters.[66] In late May, James Bell submitted a five-page report to Lund in which he discussed the quartz procurement situation. In his opinion, all of the interested agencies (Signal Corps, MRC, and Board of Economic Warfare) were unhappy with the current situation, yet no one offered to do anything about it.[67]

He felt that the policy of delaying 20% of the purchase price until after NBS inspection was a complete failure. The long delays resulted in the Brazilian exporters getting their complete payment before any part of the shipment had been inspected. This, in Bell's opinion, led to a loss of respect for the U.S. government on the part of the exporters and the Brazilian government. The fact that some shipments had been ultimately discovered to contain as low as 3% usable quartz served as evidence for this conclusion. The allegations of fraud and scandal only served to further weaken the position of the MRC program. Furthermore, it appeared to Bell, from reports by civilian importers, that the policies of the MRC had actually *decreased* the amount of high-grade quartz (as the miners were able to make just as much money from the low-grade material as they could from the much more labor-intensive high-quality crystals). Even though the program was adding needed quartz to the MRC stockpile, it was doing so at too high a monetary cost and at a loss of respect on the part of the Brazilians, the American importers, and the crystal manufacturers. Stating that "the Metals Reserve Company should be run on a satisfied customer basis," Bell urged Lund to hold a conference of all interested agencies and attempt to solve the current problems.[68]

The result of Bell's report and the previous months' worth of warnings and complaints was that the U.S. and British governments entered into a new round of negotiations with the Brazilian government regarding the prices and quality of raw quartz. While these negotiations were going on, the MRC halted all further purchases of raw quartz. This buying halt continued until the fall.[69] It was hoped that by stepping out of the market for a short period of time, the price situation would return to what it was at the beginning of the year. This was not the case, however. It appears that the U.S. departure left the market flooded with low-quality quartz, resulting in "strenuous competition" between the private importers for the small amount of high-quality quartz available, thus driving prices even higher.[70]

On August 22, 1942, a meeting was held between representatives of the WPB, the MRC, the Board of Economic Warfare, and the British Ministry of Supply. One option discussed for stabilizing the quartz market was the use of the WPB's General Imports Order M-63.[71] By placing quartz under the authority of M-63, the WPB would assume control over all aspects of quartz procurement, including mandating prices for both government and private importers, the licensing of private importers, and the oversight of all raw quartz distribution to industry (from both government and private stocks).[72]

Though negotiations between the governments continued through the summer and into the fall, raw quartz was piling up in Rio due to the lack of MRC purchases. On an inspection visit in July, Bell found that nearly 1,500 tons of quartz was sitting in exporters' warehouses in Rio. Not more than 20%, in his estimation, would likely pass the current 30% usable-volume specification for MRC purchase, however. Another meeting was held in Washington in early September in which new purchasing procedures were recommended. Quartz should be added to the list of M-63–controlled commodities, but 100% of the contract price should be paid up front, provided 100% of the quartz was inspected before shipment.[73]

In October, Bell returned from Rio to report that the decrease in purchases in Rio was beginning to result in a decrease in output from the mining regions. Furthermore, the new stockpile recommendations of 900,000 lbs and urgent requests from the British and Canadians served to increase the need to resume MRC purchases of raw quartz.[74] Quartz was thus placed under M-63 and arrangements were made to relieve the overstocked condition of the Rio exporters.[75] Kenneth Murray, brother and business partner of Donald Murray, was invited to Washington for discussions related to purchasing the accumulated stocks in Rio. Murray felt that his company could handle the job. His brother had estimated that he could extract the usable crystals from the lot and have them shipped to NBS within a period of eight weeks. He would also make sure that the rejected crystals were properly disposed of so that they "would not appear on the market again." Donald Murray also felt no real obligation to buy *all* the quartz from *every* exporter; he believed many of them to be mere speculators who should be "left to stew in their own juice."[76]

"WE FEEL THAT YOU ARE CONTRIBUTING A VERY VALUABLE SERVICE TO THE WAR EFFORT IN THE TRANSPORTATION OF THIS STRATEGIC RAW MATERIAL"

With the MRC back in the quartz-purchasing business, and, with demand more critical than ever, the task of quickly transporting the raw crystals into the United States assumed even greater importance. Historically, shipments of raw quartz had always made their way to the United Stated via steamships, slower, but much less expensive than the alternative of air transport. With America's entry into the war, two dangers arose that resulted in a shift from ships to airplanes as the primary method of transportation: the presence of German submarines along the coasts of North, Central, and South America, and the time required to ship the raw quartz from Brazil to the United States. The lesser of these dangers was the presence of German submarines within the shipping lanes. This menace necessitated a change in the amounts of quartz allowed to be transported on a single ship. The precipitating event for this policy change involved a Greek freighter known as the *Penelope*.

On April 21, 1942, an assistant entered R.J. Lund's office and handed him a note. The note informed him that it had recently been discovered that the *Penelope* was carrying 452 tons of quartz crystals and was currently crossing the Caribbean unescorted. It was not unusual for ships to travel alone during the early months of the war; at this time, it was still believed to be a safer means of travel than large, more easily detected convoys. Nevertheless, 452 tons of badly needed quartz crystals was simply too much to trust to fate. Lund immediately assigned his assistant, A.I. Henderson, the task of seeing what could be done to offer the ship some protection. Henderson contacted the War Shipping Administration, which passed along to the Navy a request for an escort. As a result, the *Penelope* safely completed her voyage to New York under the constant watch of coastal air patrols. After this episode, the maximum amount of quartz that would be allowed to be transported on a ship was set at 50 tons.[77] A year later, the restrictions were adjusted such that, of these 50 tons, no more than 15 tons could consist of the two highest quality grades. For ships traveling along the Pacific coasts of North and South America, the submarine menace was considered to be much less; such ships could transport up to 150 tons, no more than 50 of which could be high grade material.[78]

Though several claims of quartz ships being lost to German U-Boats have been made in memoirs and personal remembrances, the historical record does not back them up.[79] *Lloyd's War Losses*, the compendium of information on all Allied and neutral shipping losses during World War II, lists only three ships as having quartz among their cargo. All three, the Greek-registered *Granico* and the British-flagged *Fort Chilcotin* and *Baron Semple* were lost en route from Rio de Janeiro to Freetown and the United Kingdom.[80] The *Granico* was sunk by the Italian submarine *Finzi* near the West African coast, the *Fort Chilcotin* by the German U172 not far from the Brazilian coast, and the *Baron Semple* by U848 midway between the South American and African continents.[81]

The official history of the War Production Board's quartz program mentions the loss of a quartz cargo ship in the Caribbean "during the week of July 18," 1943.[82] No such ship is explicitly mentioned in *Lloyd's War Losses*, however, it is possible that its cargo was not described in detail (some ships were simply described as carrying a cargo of "ore" or "minerals"). Thirteen cargo ships are listed as sunk while traveling from Rio to the United States, carrying cargos of ores, minerals, and "general cargo." The most probable identity of this claim is the American-flag ship *African Star*, sunk July 12, 1943, with a cargo described as "chrome ore, asbestos, and other general cargo."[83] However, the *African Star* was still in the South Atlantic when sunk, far south of the Caribbean.[84] It is possible that some quartz was lost on Brazilian-flag ships. Thirty percent of the coastal shipping fleet of Brazil was lost to German submarines during the war, a great many of which show no cargo information in *Lloyd's War Losses*.[85]

The greater of the two dangers inherent in transporting raw quartz from Brazil was time. A steamship could take two to three months to reach its destination.[86] By the fall of 1942, the industry could not wait this long for raw materials. Throughout the remainder of the war, the majority of the quartz transported from Brazil would travel by airplane.[87] Some private importers had already been chartering airplanes for transporting their shipments. Pan-American Airways was the primary carrier between the United States and South America, though other smaller airlines were utilized. A letter dated June 9, 1942, exists in the files of the OCSigO written by O'Connell to the owner of Taca Air Lines of New York City. In his letter, O'Connell thanks the airline for transporting quartz for the Donald Murray Company "on several occasions." He points out the "valuable service to the war effort" of transporting this "strategic material" and expresses his hope that they will continue such service.[88] After the October decision to have the MRC resume their buying, there would be plenty of business for any and all airlines willing to transport quartz.

Most of the air transportation burden would be born by the Air Transport Division of the Board of Economic Warfare. With ocean transport being too slow, and the preference of Pan-American for the much more profitable passenger service, the WPB called upon the Board to bring its resources to bear to quickly bring the purchased quartz into the country.[89] Utilizing Army and Navy cargo planes, both MRC and private importers' stocks could be flown from Rio to Miami in two days.[90] Whereas the MRC incurred no expense, private importers were charged $1.50 per pound for the service.[91] The Army eventually realized that this transportation fee, charged by the MRC for the use of *Army* aircraft, was being passed on to it by the crystal manufacturers (who had had it passed along to them by the importers). Though the Army suggested that these funds be passed along to it by the MRC, there is no evidence that such an arrangement was ever carried out.[92]

By mid-October, Donald Murray had completed his sorting of the quartz stocks on hand in Rio and had eight tons ready for shipment. Rather than wait for the next available cargo ship, the State Department cabled the U.S. Embassy in Rio to make arrangements for the MRC to take temporary ownership of the quartz, expediting its transport by airplane. The shipments would be transferred back to Murray in the U.S. and an "equitable service charge" would be assessed.[93] Within a year, the supply situation would have improved enough that the priority assigned to quartz for air transport would be decreased. However, until that time, nearly one million pounds of MRC-owned raw quartz would be flown to the United States along with nearly one-half million pounds of private importer stocks.[94]

"ONE SMALL CONSUMER IS COMPLAINING BITTERLY ABOUT THE QUALITY OF THE MATERIAL WHICH HE RECEIVED"

Of the raw quartz imported during 1942 and 1943, 60% came into the country through private importers. This material would be sent directly to the contracting manufacturer. Quartz imported through the efforts of the MRC went first to the NBS for inspection and then was stored as part of the national stockpile. Private importers, however, were not able to keep up with the industry's demand. Approximately 20% of the industry's needs were supplied with quartz from the MRC stockpile; the MRC's holdings could be tapped much easier than those of the Treasury Department (which, as will be discussed below, required an Executive Order from the President to be distributed).[95]

Though it was easier for manufacturers to get raw quartz from the MRC stockpile, it still was not a simple or completely satisfactory procedure. Requests for quartz from the MRC stockpile had to first be approved by R.J. Lund's office at the WPB. Manufacturers were requested to supply detailed information on their operations, including "consumption of quartz over the previous three months, production of plates, names of customers with priority ratings, and present stocks of raw quartz on hand." Furthermore, the WPB was interested in estimates of future production.[96] Requests approved by the WPB were forwarded to the MRC, which would allocate the needed amount and authorize Leonard Buck to complete the sale.[97] Quartz from the stockpile was classified only according to weight class and usable volume. No information regarding dimensions or presence of crystal faces was given. Furthermore, lots would be sold as "general assortments" of all sizes and usable-volume classes.[98]

Initially, a great deal of apprehension as to the overall quality of the stockpile existed on the part of both manufacturers and the government. In order to satisfy its own curiosity, Bendix Radio wrote to Lund two weeks after Pearl Harbor requesting a test sample of MRC quartz.[99] Lund felt that this was an excellent idea and authorized the shipment provided that Bendix would later submit a detailed report of the amounts of plates that were manufactured from the various size and quality classes.[100] Soon thereafter, other large manufacturers (such as General Electric and Westinghouse) expressed similar interest in making test runs of stockpile material.[101] By late January 1942, the requirement of reporting on the general quality and usability of MRC quartz became standard; any company purchasing MRC material was required to submit such reports.[102]

Many of the reports came back fairly negative. The small size of the average stockpile crystal was problematic for some. Others found large fractions (some as large as 25%) of their shipments to be completely unusable (even though it had passed inspection at the NBS). In some cases, crystal manufacturers "complain[ed] bitterly about the quality of the material" received. James Bell wrote to Lund in May 1942 informing him of a general feeling of "mistrust" of the stockpile material on the part of consumers. His strongest piece of evidence for this was that "no one who has received a shipment has asked for another order." In Bell's opinion, the issue of quality, along with the inability of the program to respond quickly to emergency orders (due to the "cumbersome" procedure of going through the WPB, the MRC, and Leonard Buck's office in order to process a sale) was preventing the MRC from adequately serving the needs of the industry.[103]

Among Bell's suggestions for improving the system was to centralize the responsibility for selling stockpile quartz. Bell felt there was no clear "division of authority" between Buck and the MRC: "Negotiations are carried on with Mr. Buck's office about which the War Production Board knows nothing until Mr. Buck notifies the Metals Reserve Company that someone has requested to purchase a certain amount of crystal." Bell felt that authority over all facets of the process should be centralized in Washington. Furthermore, if the quality of the stockpile quartz was indeed less than that available from private importers, Bell felt that it should be priced accordingly.[104] As will be discussed in detail in Chapter 8, the issue of quality was in reality a problem of perception on the part of manufacturers. During the first year of the war, as the industry learned to produce crystal oscillators in larger and larger quantities, a great deal was learned about what types of crystals truly were usable for oscillator production. As the industry learned to deal with what had earlier been considered "reject" material, the complaints regarding the stockpile quartz greatly diminished.

"IT HAS COME TO THE ATTENTION OF THIS OFFICE THAT THE MONITOR PIEZO COMPANY IS NOW CLOSED DOWN BECAUSE THEY HAVE NO QUARTZ . . ."

Companies requesting stockpile quartz could not specify the sizes or types of crystals they wanted. Though they could inspect the lots selected for them at the NBS (chosen to represent a cross-section of the stockpile material), they could only agree to accept or to reject the entire lot. Most manufacturers, though they may have been less than enthusiastic regarding the quality of the material, were able to utilize it, for the most part, in their plants. One company, however, due to its own decisions in setting up its plant, was not able to utilize the stockpile quartz, leaving it vulnerable to the periodic dips in supply. As discussed in Chapter 4, Herbert Blasier designed his production line at Monitor Piezo to utilize a single type of crystal (long, thin, faced crystals known as "candles"). Blasier depended on private importers, primarily Donald Murray, to supply his raw quartz. At times, however, his supply situation became sufficiently critical that direct intervention on the part of the Signal Corps or the government was required to keep Monitor in operation.

In early March of 1942, Monitor was having trouble getting its shipments on time from Donald Murray. Blasier contacted Lund about the possibility of purchasing quartz from the MRC stockpile, stressing his particular needs with respect to crystal size and shape. Lund could not meet these specific needs, but did authorize shipment of 1,000 pounds of medium-sized crystals "having a minimum usability of 50%."[105] This shipment did little to meet Monitor's needs, however, and their supply situation became more serious. By April, Monitor was very close to having to shut down production due to lack of quartz. The decision was made to request a release from the Treasury stockpile; much higher-grade material that should contain adequate amounts of the type of crystal needed by Monitor. Such a release, however, required the direct approval of the President. Thus, over Easter weekend, an Executive Order was drafted authorizing the release of 100 pounds of crystals from the Treasury stockpile. Nothing could be done, however, until the Order was signed. The document sat on the President's desk all day Monday. By Tuesday morning, April 7, anxious telephone conversations were being held between representatives of the WPB and the Signal Corps Quartz Crystal Section. Later that afternoon, Roosevelt finally signed Executive Order 9123, and Lund and Bates personally selected the crystals for Monitor, shipping them out the following day.[106]

Though such efforts were never required again on Monitor's behalf, the company did have frequent problems with supply throughout the first year of the war. Though the Donald Murray Company appeared to be procuring adequate amounts of the quartz "candles," frequent problems arose in transit. The worst occurred in late August when the company actually had to shut down briefly. The Signal Corps was at a loss to understand how this could have happened. The Quartz Crystal Section itself had worked to guarantee high priorities for the shipment of Monitor quartz on both Pan-American and Taca Airlines. How, they demanded of the WPB, could such a situation have occurred? The WPB solved the problem, however. A small sample of Treasury quartz, sent to the Treasury office in Los Angeles earlier in the year as insurance against just such a situation, quickly got the plant back into production (and resulted in a very gracious thank you letter to Lund from the Quartz Crystal Section's Major Olsen).[107]

"THE BRITISH MAINTAINED IT WAS ABSOLUTELY IMPOSSIBLE FOR THEIR INDUSTRY TO USE ANY LOW QUALITY MATERIAL WHATSOEVER"

Unfortunately, no such quick fixes existed for the problems of the WPB's other large and somewhat troublesome customer, Great Britain. Just as the Quartz Crystal Section worked closely with representatives of the British crystal industry attempting to increase its production capacity, the WPB struggled throughout the war in an attempt to supply the British with the amount, and quality, of quartz that they required. In many ways, the problems of the British industry were similar to those of Monitor Piezo: a stubborn dependence on large, high-grade crystals that had never been in large supply even before the expansion of the American industry.

America's large-scale entry into the Brazilian quartz market resulted in a dramatic decrease in the amount of quartz available to the British, coupled with increases in price. By the summer of 1942, the British were anxious to enter into a purchasing agreement with the U.S. government that would result in its becoming the sole buyer of quartz in Brazil (with the United Kingdom purchasing its quartz directly from U.S. stocks). This arrangement was not viewed favorably by American officials, however, as it would leave all buying, inspecting, and shipping responsibilities to the United States. Furthermore, most of the highest-quality crystals would be lost to the British, leaving primarily low-grade material for the United States. Though eager to assist the British, the Americans did not agree to the proposed plan.[108]

In the opinion of the U.S. government, a great deal of improvement in the British supply situation could be had if they would simply convert their crystal industry to the use of smaller, lower-quality crystals. Clifford Frondel attempted to impress this idea upon them during his time there. Also, X-ray equipment was sent to Great Britain to aid in the orientation of smaller, unfaced crystals. These efforts were unsuccessful, however. The British felt it would take too much time, resources, and personnel to make the changes the Americans were suggesting; three items that were already in short supply. They clung to their demands for large (one pound or larger) pyramidal crystals throughout the war.[109]

Another commodity the British were short of was money. In agreement with the private importers, the British purchasing agents felt that the buying practices of the American agents were driving up the price of quartz dramatically. On the other hand, James Bell of the WPB felt that the British were loath to pay anywhere near fair prices for their quartz in the first place. A less-than-happy relationship developed between the British and U.S. purchasing commissions in Rio. Each accused the other of paying too high prices for low-quality quartz. This last accusation actually seems to have been true of the British. Whereas the Americans had developed a fairly detailed inspection process utilizing oil baths and polarized light sources, the British method of inspection consisted primarily of naked-eye observations. To test the accuracy of the British methods, and hopefully to settle the ongoing arguments, a sample of British-purchased quartz was shipped to the NBS for inspection utilizing the American methods. It turned out that only 15–25% of what had been claimed by the British to be "grade A" material was actually high-quality quartz. As much as 45% of the inspected lot was considered by the Americans as very low-grade quartz. The British continued to purchase and inspect their own lots of crystals throughout the duration of the war. Though it would be suggested several times, no official cooperative buying agreements were ever enacted. The United States did, on several occasions, send the British emergency shipments of quartz and facilitate the air transport of British quartz purchases from Brazil to Great Britain.[110]

The quartz supply business, whether supporting the domestic industry or foreign allies, was a complicated undertaking. Agencies of the civilian government and the armed services struggled throughout the war with difficulties related to price, quality, transport, inspection, and distribution. Just as no one really knew how to go about mass producing crystal oscillators prior to the war, no one knew much about procuring raw materials from foreign countries in the vast amounts needed by wartime industry. Whereas a great deal of the necessary work on the production side of the issue was carried out by a rather small group of people within the Signal Corps, likewise a fairly small group of civilians on the scene in Rio and the interior regions of Brazil deserve a great deal of credit for getting the U.S. purchasing program on track and alleviating many of the problems that plagued it during the first year of the war. The story of these people is the subject of the following chapter.

1. An open-pit quartz mine in the Sete Lagoas District of Brazil. 169-GA-6B-6

ILLUSTRATION CREDITS:
Unless stated otherwise, all photographs are from the National Archives Still Pictures Collection. Archival identification numbers are given for each photograph.

2. A close-up of workers in a quartz mine in the Sete Lagoas District of Brazil. 169-GA-6B-25

3. Hand sorting of quartz crystals. Rio de Janeiro, Brazil. 169-GA-6B-26

4. Checking quartz crystals for internal flaws utilizing a mineral oil bath. Rio de Janeiro, Brazil. 169-GA-6B-2

5. Crates of quartz crystals awaiting shipment to the United States. Rio de Janeiro, Brazil. 169-GA-6B-16

6. A U.S. sailor takes a nap with his head cradled on a load of Brazilian quartz. He is riding on a Naval Air Transport plane, northbound to the United States. 169-GA-6B-15

7. Robert D. Butler, Bureau of Economic Warfare representative. Rio de Janeiro, Brazil. 169-GA-6B-17

8. Julian Mateson, Herbert Jelinek, and Fred Johnson (left to right) of the Foreign Economic Administration. Sao Salvador, Brazil. 169-GA-6B-31

9. A Signal Corps technician places a set of quartz blocks onto a lapping machine, the initial step in the crystal grinding process. Philippines, 1945. SC235281

10. A member of the Crystal Grinding Section, 832nd Signal Services Co. checks the frequency of quartz blanks after initial lapping. Australia, 1944. SC268431

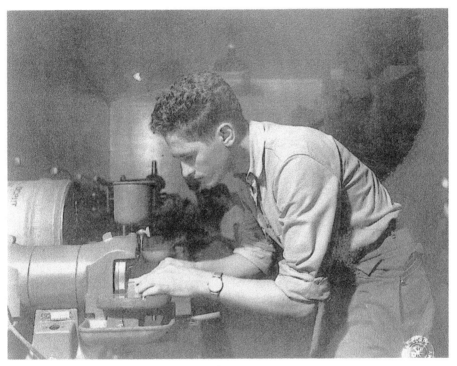

11. A Signal Corps technician carries out edge grinding on a set of quartz blanks. Philippines, 1945. SC235280

12. Hand grinding, the final stage of the production process, is carried out by a Signal Corps technician. Philippines, 1945. SC235279

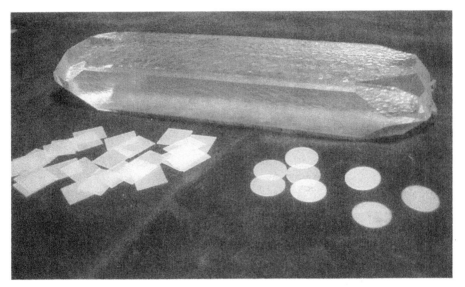

13. A bar of cultured quartz alongside examples of the types of wafers that might be sawed from it. (The bar is 20.5 cm long and 2.5 cm thick; the blanks are only 3.5 mm thick.) Photo by the author; originally published in *QST Magazine*, January 2004

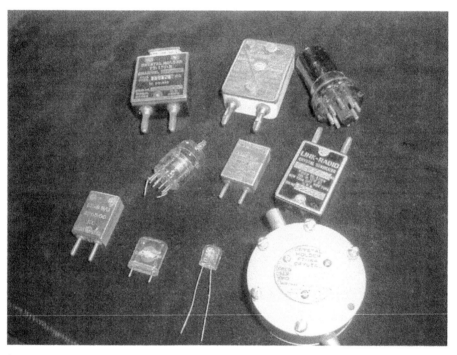

14. Examples of the types of holders used in crystal oscillator units. Photo by the author; originally published in *QST Magazine*, January 2004

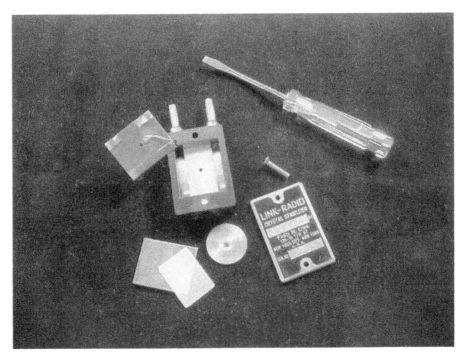

15. A common crystal unit shown disassembled. Notice the quartz wafer at bottom center. The metal plates and disk serve as electrical contacts. Photo by the author; originally published in *QST Magazine*, January 2004

16. Major General Roger B. Colton (left) and 1st Lt. Edward E. Ragon. Pentagon, 1943. SC169077

17. Radio operator seen at his radio that is permanently mounted in a jeep. SC140175

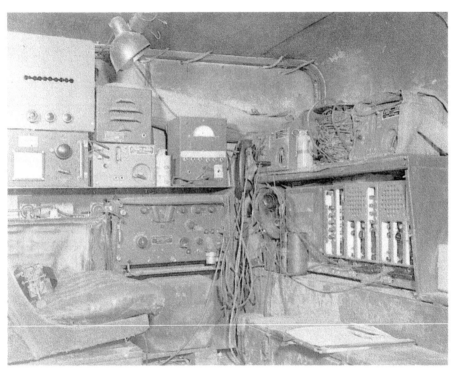

18. Interior of a radio truck of the 69th Infantry Division. Belgium, 1945. Notice the leg-mounted telegraph key on the seat. SC392816

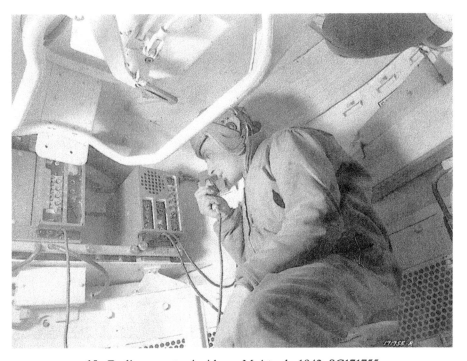

19. Radio operator inside an M-4 tank. 1943. SC171755

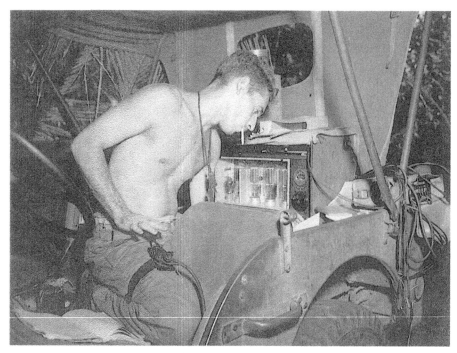

20. Radio operator (with leg-mounted telegraph key) using a radio set temporarily set up in a jeep after original position was bombed. New Guinea, 1943. SC178204

21. Field radio set operating near airport. Maison Blanche, Algeria, 1943. SC167433

22. Radioman of the 99th Division, U.S. 1st Army takes shelter in roadside ditch during heavy German shelling on outskirts of Honningen, Germany, 1945. SC395799

23. Radio operator contacts the rear to report on the progress of patrol of 10th Mountain Division on the Apennines mountains in Italy, 1945. SC276937

24. Two riflemen fire at a cave while a radio operator relays progress. Okinawa, 1945. SC272384

25. Shrapnel holes give evidence of just how close to the action the radios often came. North Africa, 1943. SC166598

7

"THE WHOLE RADIO CRYSTAL PROGRAM OF THE ARMED SERVICES DEPENDS UPON THE SUCCESS OF THE PROCUREMENT PROGRAM IN BRAZIL. NOTHING MUST BE ALLOWED TO INTERFERE WITH IT"

As the sun rose over the mountains to the east, the quartz miners would leave their thatched-roofed huts, grab their picks and shovels, and lead their donkeys down to the mines to begin another day. Down in the open pits, the miners would spend their day chipping away at the terraces that covered the sides of the mines. As the quartz occurred in long-running veins as well as isolated pockets, this was the most efficient method of locating and extracting the crystals.[1] It was back-breaking work; for every few pounds of quartz uncovered, hundreds of pounds of dirt, rock, and unusable "milky" or "bull" quartz would be carted out of the pits by the donkeys. At the end of the day, the workers returned to their huts while the music of homemade guitars filtered across the camps. These events took place day after day in dozens of locations across the interior regions of Brazil. An estimated 50,000 miners took part in this work over the course of the war, making very little in the form of wages and being left precious few other opportunities after the war when the industry returned to its pre-war levels.[2]

As primitive as the operations were, they were to be considered industrial compared with the mining methods just a decade earlier. In the pre-war years, garimpeiros ("quartz hunters") would work crystals from surface deposits or from small mines, collecting enough to trade to local merchants for necessities (it has been said that increases in the price of quartz could actually result in

Crystal Clear: The Struggle for Reliable Communications Technology in World War II,
by Richard J. Thompson, Jr.
Copyright © 2007 by Institute of Electrical and Electronics Engineers

decreases in the quantities mined, as the miners could now get their needed supplies by trading fewer crystals).[3] The crystals might change hands several times before completing their several-hundred-mile journey to Rio de Janeiro.[4]

With the increase in demand for quartz in the late 1930s, the mining operations increased in size, though very little in technological sophistication. Mining operations were concentrated in four elongated regions or belts, fairly parallel to each other (Figure 7.1). Approximately half of the production

Figure 7.1. Map of the quartz mining regions of Brazil

came from the Minas-Bahia Belt stretching across the states of Minas Gerais and Bahia. Another 30–40% came from the belts within the western Goias region. The remainder came from a small region known as the Coastal Belt, located within the state of Espirito Santo. Individual deposits were by no means extensive; most of the mines produced at most 500 pounds of clear, radio-grade quartz each month.[5] Furthermore, with the poor conditions of the few roads across the interior provinces of Brazil, it remained a time-consuming effort to transport the crystals to the exporters' warehouses in Rio.

"THE PURCHASING PROGRAM TO DATE HAS NOT BEEN SATISFACTORY AND HAS RESULTED IN A STATE OF UNEASINESS IN THE RIO MARKET . . ."

Whereas the conditions under which the quartz was mined and transported could be described as primitive, the minerals market in Rio de Janeiro was as modern and "chaotic" as that of any other large city. The U.S. government entered the quartz market in 1940, in direct competition with both private importers and foreign governments, with very little knowledge or understanding as to how the Brazilian quartz trade worked. It soon became critical that the government's quartz purchasing agencies put some representatives in place in Rio to learn what they needed to know to succeed in this new market.

The first such "fact finding" mission took place in January 1941. Stephen Capps of the U.S. Geological Survey led an inspection tour of the mining region near Sete Lagoas (approximately 300 miles northwest of Rio). It was on this trip that the idea of supporting the mining operations by purchasing significant amounts of low-grade material first surfaced. Warned by the mine owners that the loss of the Japanese market for their lesser-quality crystals could result in their shutting down operations entirely, Capps passed the suggestion along to the MRC.[6]

Making the market situation even more cloudy, the Brazilian government entered the picture in February 1941 with their declaration of a 10% *ad valorem* export tax. In order to maintain the price of quartz, the government instituted a system of price floors or "tabellas," minimum prices below which quartz could not be purchased for export. As part of the tabella decree, a system of quality specifications was produced, dividing quartz into either high grade ("A"), medium ("B"), or low grade ("C"). This represented the first attempt at producing a set of quality classes for raw quartz. The initial tabella prices set were well in excess of the current market prices. Though not looked upon favorably at the time by the importers or the U.S. government, the price structure did improve the financial state of the mining enterprises, keeping them in production throughout the war.[7]

Three months later, C.B. Hamilton was sent to Rio to serve as the purchasing agent for the MRC. He instituted the purchasing policies that were met with such disapproval from the private importers (as discussed in Chapter 6). In a letter similar to that of Donald Murray quoted in the previous chapter, Hamilton was referred to by a Rio-based importer as "being certainly 100% honest," but having "made a fool of himself." Hamilton returned to the United States at the end of the year.[8]

A third attempt to get a handle on the quartz market from an on-site inspection was initiated in July 1942, when James Bell left for Rio, leaving Robert McCormick as Acting Chief of the WPB Quartz Crystal Section. Intending to stay for only three months, Bell actually stayed for a year, taking over leadership of the entire Brazilian operation for the Board of Economic Warfare. Bell was instrumental in getting the MRC purchasing program back into operation after the six-month hiatus of mid-1942. In his words, "The purchasing program to date has not been satisfactory and has resulted in a state of uneasiness in the Rio market, which in turn is beginning to slow up production in the interior."[9] Resuming quartz purchases would do a great deal to alleviate the "uneasiness" in the market. However, one of the requirements for the resumption of government purchases was that 100% of the quartz would be inspected before it ever left Rio de Janeiro. This would require the establishment of a full-scale inspection program under the auspices of the Board of Economic Warfare.

"THE MEMBERS OF THE SIGNAL CORPS CONTINGENT . . . CONTRIBUTED SIGNIFICANTLY TO THE PROGRESS OF THE U.S. PUBLIC PURCHASE PROGRAM"

Establishing a crystal inspection facility in Rio was not a new idea. As early as May 1941, the Signal Corps was being advised as to the benefits of inspecting quartz prior to shipment.[10] No serious efforts toward this end were made, however, until the fall of 1942. In September, Major Olsen of the Quartz Crystal Section began a steady correspondence with Major Horace Wood, director of the General Development Laboratory at Ft. Monmouth, regarding the selection of a group of men to be assigned to the Rio inspection unit. Six men were requested to be sent to Rio immediately, with another dozen or so to follow in November.[11] The inspectors would be assigned to the MRC, which would pay their salaries and take care of the arrangements for transporting them. As some of the men initially approached for this duty turned down the assignment due to the relatively low wages offered (a maximum of $2,500 per year with no per diem allowance), Olsen suggested to Wood that he seek out unmarried men who might be more willing to accept such a salary.[12] Of the 100 candidates for the positions, 18 were ultimately accepted. By late September, the first group of men had been chosen and were undergoing the various procedures required of foreign service (such as draft releases and vaccinations).[13]

The first six men, Louis Landrau, Usher Mayer, Louis Rudovsky, Alexander Saldan, Nicholas Colon, and Tirso Mattei, left Ft. Monmouth during the third week in October. They received their passports and travel clearances from the OCSigO and then proceeded by train to Miami. From Miami, a Pan-American plane would take them on to Rio. By October 30, all six had reported for duty to the U.S. Purchasing Commission (the official designation of the Board of Economic Warfare's Brazilian program).[14] The remaining dozen men (including two released from duty at the NBS inspection facility) arrived over the course of the following three weeks.[15] One week before the first group departed from Ft. Monmouth, Colonel O'Connell wrote to Colonel Messer at the American Embassy in Rio informing him of the imminent arrival of the Signal Corps men. Expressing his concern for the civilian employees being assigned to Rio without a Signal Corps officer to oversee their efforts, he asked Messer to "take a fatherly interest in the boys and tell them how to keep out of trouble." From the fact that three members of the unit were returned to the United States within weeks (one for "unsatisfactory service," one for "insubordination," and one, Alexander Saldan, for "mutiny"), it seems that O'Connell's fears were justified.[16]

Upon arrival, the inspectors began setting up their facilities. Their efforts were initially hampered by a lack of equipment and a generally poor location for their laboratory (the basement garage of a large office building. While waiting for additional equipment to arrive (much of it being designed and built at Signal Corps depots), inspection was organized into two shifts to better utilize the equipment available.[17] By January, the group had moved into a much more suitable facility adjacent to the American Embassy and nearly 70 tons of quartz had been inspected.[18] During this time, Julian Mateson and Sidney Sandler, the former NBS employees, were designated as day and night foremen, respectively. Clarence Wynn was named as chief inspector. Wynn was also the chief instructor, coordinating the training of local civilians with the intent of eventually replacing the much higher paid American inspectors.[19] Wynn also trained representatives from several quartz exporters and a team of American mining engineers, along with a few inspectors from the Brazilian and British governments. In all, over 400 people received training as quartz inspectors.[20]

In January, the first instances of discipline problems among the Signal Corps inspectors occurred. The MRC had expressed concern with some of the men selected as early as the previous September. "Some are married and want to take wives, some are possible trouble makers and complainers."[21] Along with the three men mentioned above, a fourth was sent home in January for an "alleged" stomach ailment.[22] One final inspector, Wesley Clifford, was sent home early, though he was transferred to the NBS, where he provided good service.[23] It seems that Clifford's wife back in the United States was sick and he missed a great deal of work because he was worried about her. When it was discovered that his favorite place to do his worrying was on the Copacabana Beach, arrangements were made for his transfer.[24] With the exception of these men, the remainder of the inspectors provided adequate service. By April 1943, however, enough locals had been trained that all but five of them were ordered back to the United States. The remaining inspectors were Wynn, Mateson, Landrau, Stanley Mitchell, and Arthur Slimm.[25]

These men assumed leadership roles in various components of a rapidly expanding program. In January, Mateson had been transferred 900 miles up the coast to the city of São Salvador, Bahia, to set up an inspection station there. Slimm joined him in March and the laboratory went into full-scale production in May. That same month Mitchell was placed in charge of a new warehouse facility at which rejected quartz underwent salvage attempts (by trimming away portions with defects or electrical twinning). Later that summer, Mateson's organizational skills were called upon once more in an effort to establish an inspection station closer to the mines in the northern regions of the country. Though efforts were made to divert quartz to the new station at Belém, Pará, very little was received. The stocks at the Bahia station were rapidly increasing, however, and Mateson was ordered in October to close the Belém station and return to Bahia. Meanwhile, Wynn and Landrau continued their training efforts at the Rio lab.[26]

In November, Arthur Slimm became ill and was transferred back to Rio; he was replaced in Bahia by Louis Landrau. Mateson's responsibilities increased rapidly; by January 1944, he was given authority over most of the commercial activities of the inspection lab, including control of the local bank account. Slimm, it seems, became unhappy with his duties in Rio and requested a transfer back to the United States. Not wanting to lose a man of his abilities, however, Wynn assigned him to work with Mr. E.E.S. Brown, the British purchasing agent. Slimm was a great help to the British in their inspection efforts and carried out a study of the differences between the British and U.S. inspection procedures.[27]

Through the fall of 1944, Wynn continued his educational duties while overseeing the work at the Rio and the Bahia labs. He also spearheaded the quality control efforts of the two inspection labs, maintaining consistent inspection standards. After concluding his work with the British purchasing agency, Slimm was transferred to Sete Lagoas to serve as a field technician. By the end of the year, however, a back condition (either due to or aggravated by riding in jeeps over rough terrain) forced his transfer back to the United States. His duties in the interior were taken over by Mitchell.[28]

Robert Butler, the chief of the U.S. quartz program in Brazil, credits the Signal Corps inspectors, particularly the final five, with "contributing significantly" to the overall success of the purchasing program.[29] Their ability to rapidly work through the tons of quartz that had accumulated in Rio during the U.S. purchasing hiatus, along with their training of hundreds of local inspectors, were of paramount importance to getting the flow of quartz into the country back on track. During their first four months of work alone, over 700 metric tons of quartz (with a value of over $2 million) was purchased and sent on its way to the plants of the U.S. crystal industry. Nearly 1,500 tons were inspected and purchased through the end of the war.[30]

Though their living conditions were at times less than desirable, and frequent problems arose concerning the payment of salaries and travel expenses, the large majority of the Signal Corps men stationed in Brazil carried out their duties conscientiously and successfully. Complete inspection before purchase was an essential element of a fair and efficient raw quartz purchasing program. The Signal Corps inspectors guaranteed such a program.

The Brazilian inspection and purchasing programs were major efforts, involving the Signal Corps, the Metals Reserve Company, and the Board of Economic Warfare (later renamed the Foreign Economic Administration). The close cooperation between these various agencies contributed greatly to their success. Another major effort designed to increase the flow of quartz from Brazil was a program aimed at improving the production of the mining operations. This program would likewise involve several government agencies, including the Signal Corps, the War Production Board, and the Reconstruction Finance Corporation.

"WE HAVE BEEN ADVISED THAT THE USE OF MACHINERY WILL GREATLY INCREASE THE OUTPUT AND DELIVERY OF BRAZIL QUARTZ"

Dick Stoiber had never been on an airplane before the day that he and Lieutenant Colonel Herbert Messer left Washington for Rio de Janeiro. As his wife watched the plane taxi down the runway, she wondered if she'd ever see him again. He entertained similar thoughts.[31] Messer had taken charge of the Quartz Crystal Section in February 1943. Due to his experience as a military attaché in Rio, General Colton also appointed him "Special Representative of the Chief Signal Officer for liaison with the U.S. Purchasing Commission" in Brazil. Instructed to make such trips to Brazil as he felt necessary, Messer was also advised to take Dick Stoiber along whenever possible. Stoiber's expertise in mining geology was considered very valuable to the quartz program in Brazil.[32]

Their DC-3 took off from Bolling Field in Washington and carried them to Miami. Day two of their trip took them from Miami to Trinidad. After three days of dawn-to-dusk travel (being unable to fly at night due to blackout conditions), they arrived in Rio.[33] The primary purpose of this trip was for Messer and Stoiber to serve as advance agents for a new program aimed at increasing mining production through mechanization. Nearly $1 million of earth-moving equipment was on its way from the United States to the Brazilian interior and Stoiber needed to determine where and how it should be put to use.[34]

In late November and early December of 1942, a great deal of uncertainty existed within the U.S. government as to the ability of the Brazilian quartz mines to supply the needs of the crystal industry. Even though the Signal Corps inspectors were now on the scene in Rio, no one knew how quickly they could begin clearing quartz for shipment to the United States. In order to increase the overall supply of raw quartz, the decision was made within the Board of Economic Warfare (BEW) to attempt to increase the production of the quartz mines through mechanization. On December 5, Arthur Paul of the BEW's Office of Imports wrote to General Colton requesting Signal Corps assistance in purchasing and transporting the needed equipment. As it was believed that the Army could procure the equipment much faster than the BEW, a plan was proposed in which the Signal Corps would acquire the bulldozers, tractors, and other heavy machinery, and the MRC would in turn purchase them from the Army.[35]

Two days later, a meeting of representatives of the OCSigO, the WPB, the MRC, and the BEW was held. Lieutenant Colonel Zermuehlen informed everyone of the MRC's suggestion for purchasing the equipment, but expressed the Signal Corps' opinion that, as very little of the equipment was ever utilized by the Signal Corps, the MRC could most likely acquire it as easily as they could. To expedite the process, however, a letter would be written to Donald Nelson by General Brehon Somervell (Commanding General of the Army Service Forces, the branch of the Army encompassing the Signal Corps) requesting him to personally oversee the assigning of priorities and permissions for the purchases. It was felt that this would guarantee the efficient procurement of the needed machinery.[36]

The MRC representatives agreed to take on the task of purchasing and transporting the equipment, provided such assistance from the WPB was given. The letter was forwarded to Somervell for his signature one week later. In a cover letter, Colton summarized the current quartz supply situation, pointing out that "the present rate of delivery of raw quartz in Brazil is less than 50% of the current rate of use" in the United States. He also noted that several government agencies had advised him that "the use of machinery will greatly increase the output and delivery" of quartz from Brazil. Thus, in order to establish adequate priorities for the purchases, the accompanying letter to Nelson was considered essential.[37]

The actual letter to Nelson was fairly brief, making note of the MRC's request for assistance, the benefit that would be gained from Nelson's personal attention, and an acknowledgement of "the splendid work which you people have already done in supplying our needs for quartz." Attached was a detailed spreadsheet of the one million dollars' worth of equipment needed, along with information on when and where it would ultimately be utilized. The spreadsheet was divided into two sections: one for "Equipment" and the other for "Essential Supplies." The equipment included dragline shovels, caterpillar tractors, bulldozers, compressors, pumps, drills, welding and cutting tools, generators, trucks, jeeps, radios, camp equipment, and shotguns, carbines, and pistols. The essential supplies consisted of jackhammers, spikes, picks and shovels, explosives, blasting caps, and ammunition. The materials were designated to be delivered to four major regions within Brazil: Central Minas Gerais, Southern Goiaz, the São Francisco River, and the Tocantins River. The Tocantins River region was slated to receive only "essential supplies"; a note on the spreadsheet where the "equipment" entries would have gone read: "On account of dangers in this region from sickness and savages, no program contemplated except the furnishing of small tools as needed. Activities will be limited to subsidizing purchasing agencies."[38] Though the primary responsibility for acquiring the needed materials for the Brazilian program fell to the MRC, the Signal Corps did keep a close eye on the process, assisting where it could with issues of purchasing priorities and the diversion of needed equipment from other Signal Corps orders to the Brazil program.[39]

By April 1943, two-thirds of the material had reached Brazil and been delivered to the interior mining regions. The remainder (less approximately 8%, which was lost at sea) was currently in transit. With the approaching dry season, hopes were high that the equipment would result in a significant increase in mining output.[40] As it turns out, however, the mechanization efforts had only limited success in increasing the supply of raw quartz available to the American crystal industry (much less than the conservation efforts, which will be discussed in the following chapter). Explanations for the limited success of this program ranged from "the peculiar geological nature of the quartz deposits" to the "social and economic conditions of life and trade in Brazil."[41] Eventually, a large proportion of the heavy equipment was transferred to other regions and utilized in the mining of mica and other needed minerals.[42]

The quartz mining production was still very much a concern, however. By the spring of 1944, quartz purchases in Rio had declined sufficiently to set off a great deal of activity among the Signal Corps, the WPB, and the other quartz-related government agencies, each agency writing the others asking for information "concerning the reserves of quartz crystal in Brazil."[43] Two questions were of paramount importance: just what was the state of the quartz reserves and how could increased production be stimulated? The first question was very difficult to answer. Those on the scene in Brazil wrote frequently of the "extreme difficulty" of estimating the amounts of quartz remaining in the mines.[44] Although by the spring of 1945 it would become quite obvious that the mines were running out of quartz, in the spring of 1944 it wasn't even clear that decreased mining production in the interior was to blame for the drop in quartz purchasing on the coast. One of the most popular explanations was that those in the private quartz business felt the end of the war to be very near and feared the loss of any capital invested in raw quartz.[45]

Discussions were held between the Signal Corps and the Foreign Economic Administration (FEA) with regards to how to counter this trend. One

suggestion was to halt all but emergency sales of quartz from the MRC stockpile. This, it was hoped, would stimulate the private importers to increase
their purchases.[46] Though the FEA did offer some importers "protection
contracts" agreeing to purchase unsold raw quartz in the event of widespread
cancellation of private contracts, this activity was not viewed as being very
successful at stimulating production in the interior of Brazil.[47] One such
method that was attempted, along with the mechanization program, was the
loaning of funds to mine owners for improving and enlarging their mines.[48]
It appears, however, that none of these efforts were very successful at overcoming either the problems of geology or of the market place. Though both
the mining and purchasing of raw quartz continued to decline through the
remainder of the war, the earlier stockpiling efforts, combined with the
advances in conservation and utilization of raw quartz, enabled the crystal
industry to continue its steady increase in oscillator production without
serious interruption.

"QUARTZ CRYSTAL PRODUCTION ... HAS BEEN BADLY HANDLED IN BRAZIL"

The industrial buildup immediately prior to World War II was on a scale
unprecedented in all of history. The production capacity of American industry appeared to be limitless. A few public servants also saw a similar capacity
for fraud and corruption. Stating that he had "never yet found a contractor
who, if not watched, would not leave the Government holding the bag,"[49]
Senator Harry S. Truman called for the appointment of a committee to keep
an eye on the national defense industry in hopes of preventing such offenses.
Created on March 1, 1941, by Senate Resolution 71, and with Truman as its
chairman, the Special Committee to Investigate the National Defense
Program was charged with making "a full and complete study and investigation of the operation of the program for the procurement and construction of
supplies, materials, munitions, vehicles, aircraft, vessels, plants, camps, and
other articles and facilities in connection with the national defense." Throughout its seven-year existence, the committee held 432 public hearings, 300
executive sessions, and published 51 reports concerning all facets of war
production.[50]

The quartz procurement program in Brazil came under the gaze of the
committee during the final weeks of 1943. Senator Hugh Butler of Nebraska
carried out an investigation of Latin American operations, including the
quartz program. His report on the program was very negative and accusatory.
In his summary of the report, he listed nine major conclusions:

1. Quartz crystal production in Brazil had been badly handled.
2. Even though the supply of raw quartz had increased, the price paid for
   it never decreased.

3. A large and "unjustified" disparity existed between the prices paid for raw quartz in Brazil and those charged crystal manufacturers in the United States.

4. At least half of the quartz that had been imported by the MRC was useless for the manufacture of radio components.

5. Crystal manufacturers were "forced" to purchase their quartz through "a private agent."

6. The "unjustifiably high premium prices" paid by the U.S. Purchasing Commission had led to a dramatic increase in the finished products sold to the military.

7. Indiscriminate buying had led to a huge stockpile of useless quartz.

8. Graft and corruption were almost certainly involved.

9. The management of the procurement program had been "inefficient" and the sale of raw quartz to manufacturers had been "unfair."[51]

Very soon after the submission of Senator Butler's report, a response was prepared by Robert Butler, Chief of the Quartz Department of the U.S. Purchasing Commission (USPC) in Brazil. Taking each point one at a time, Robert Butler attempted to explain the circumstances surrounding each of the senator's complaints and to give what he believed to be the proper explanations. With respect to the "bad handling" charge, Robert Butler pointed out that the quartz supply had never run out and that the industry had always been properly supported. The primary drawback of the entire program, in his opinion, was the fact that Brazil had the world quartz market cornered. Furthermore, he believed that the dealers in Rio spread rumors about the USPC and tried to use political connections to increase their sales.

With respect to the increased prices, quartz was no different from any other wartime commodity. However, the prices did level off fairly soon after the start of the war. The disparity in price between that paid at the mines and that paid in Rio was due, in Robert Butler's opinion, to the mines being "a little man's business," and the export market in Rio being "a big man's business." The movement of raw quartz from the mines to the ports of Rio involved a long chain of middlemen. Though a miner might be paid relatively little per pound for his quartz, the dealers did purchase their entire lot (consisting of both good and bad crystals). At each ensuing step, the poorer crystals were culled and the prices raised in an attempt to make a profit. Thus, the prices charged in Rio were dramatically higher than those paid in the interior. Butler felt that this system was necessary in order to keep the middlemen involved in the quartz trade. Without their involvement, which demanded their turning a profit, the quartz would not make its way to the exporters' warehouses and the U.S. crystal industry would suffer.

Though some of the poorest-quality crystals were discarded as the material made its way from the interior of the country to the coast, considerable stocks of poor quality quartz did build up in Rio during 1942. The United States ultimately purchased a lot of this material in order to maintain the mining operations and to defuse an ugly political situation. Furthermore, they wanted to remove the low-quality quartz from the marketplace to prevent similar problems in the future. Robert Butler stated that this was entirely above board and that everyone concerned both knew about it and approved of it. Once the raw quartz reached the United States, manufacturers were free to purchase from whomever they pleased. The restrictions suggested in the senator's report did not exist. Also, Butler believed that, in light of the above explanation of how the quartz business in Brazil actually functioned, the prices paid by the USPC were quite justified, and that no "purchasing" of useless quartz was made as the prices for lots purchased were determined by the amount of usable quartz contained.

As could be expected, Butler took "strong exception" to the graft claim and demanded an independent and "nonpolitical" investigation be carried out with respect to the matter. Furthermore, the inspection procedures were claimed to be quite sufficient and were carried out by "loyal and honest people." There had been instances when Brazilian employees were found to be secretly working for a quartz exporter (e.g., increasing the quality rating of a lot so as to increase the price), but whenever discovered, they had been dismissed under a zero tolerance policy. The inspection of raw quartz was taken seriously by the personnel in Brazil and their work was checked after the fact by the NBS and also by the FEA.

Robert Butler felt that although there was always room for improvement in management methods, the methods of the USPC were more than adequate. It was unclear to him how judgments of these methods could be made when the judges had no basis for their conclusions; the only people who really knew of the business methods of the USPC in Brazil were the people actually working in Brazil As far as the suggestion that the manufacturers had somehow been treated unfairly throughout all of this, Butler suggested the senator ask the manufacturers themselves whether they thought the USPC had been fair or not.[52]

In light of the historical records, it seems as though the allegations of Senator Butler were largely unfounded, and that Robert Butler's response to them effectively ended the matter. If the quartz program was run as badly as he suggested, the blame would have to be shared by the Treasury Department, the War Production Board, and the Signal Corps as they all had a stake in its outcome and kept fairly close watch over it. In actuality, the Brazilian operation represents a very successful collaboration of all of these entities. The military and governmental agencies involved worked fairly efficiently from the pre-war planning stage, to the establishment of an inspection program, to the expeditious transportation of the purchased material. No one group can claim all of the credit; the success of the program lies in the fact that it was truly a team effort.[53]

8

"GOD MADE LOTS OF SMALL CRYSTALS"

Though raw quartz was making its way from the interior of Brazil to the crystal plants of the United States, it was not coming in at a rate sufficient to ease anyone's mind with respect to maintaining an adequate supply. If production could not be increased significantly, then conservation efforts would be necessary. Throughout the war, three primary types of conservation measures were attempted by the industry, the Signal Corps, and the government agencies involved: making the most efficient use of the quartz on hand, making use of quartz rejected by inspectors as being too small or possessing defects, and finding alternate sources of quartz (specifically, domestic deposits of natural quartz and the synthetic production of artificial crystals).

"OUR BIGGEST QUARTZ MINE LIES IN MORE EFFICIENT FABRICATION"

As discussed in previous chapters, the Signal Corps' Quartz Crystal Section was very interested in making efficient use of the quartz on hand. Their primary interests were identifying improved manufacturing methods and communicating these advances to the industry at large. Improved saws, X-ray orientation equipment, and the design of oscillators utilizing smaller quartz plates were just a few of the many methods of quartz conservation advocated by the QCS. The actual implementation of these techniques and pieces of

Crystal Clear: The Struggle for Reliable Communications Technology in World War II, by Richard J. Thompson, Jr.
Copyright © 2007 by Institute of Electrical and Electronics Engineers

equipment was, for the most part, up to the individual crystal manufacturers. The QCS could list certain performance-based requirements in their specifications, however, they had very little enforcement power over industry. That authority resided primarily with the War Production Board.

The WPB could enforce its rules through control of permissions and priorities for the ordering of raw materials. One particular regulation concerned with the conservation of raw quartz was Conservation Order M-146. Issued on May 18, 1942, M-146 restricted the use of quartz crystals to the production of radio oscillators and related products. Quartz could only be purchased by companies approved by the WPB and the companies were required to account for all that they possessed through the regular submission of reports on stocks, consumption, production, and shipments.[1]

The WPB agreed with the QCS on the importance of improved equipment and techniques for sawing and orientation and that oscillator plates should be made smaller where possible. Believing that the country's "biggest quartz mine lies in more efficient fabrication," they also agreed that inexperienced manufacturers, those most likely to waste large amounts of raw quartz, should be kept out of the crystal industry (or at least brought along slowly until they proved their ability to efficiently manufacture oscillators). M-146 gave them the authority to do just that.[2] Inexperienced manufacturers were not the only source of concern to the WPB and the QCS. A great deal of frustration came as a result of established crystal producers endeavoring to set up quartz cutting operations in Brazil.

Several American companies expressed interest in setting up quartz cutting operations in Brazil. The two most persistent ones were Byington & Company and American Gem & Pearl. The primary argument made by companies for establishing crystal plants in Brazil was that raw quartz cut into bars or blanks could be shipped much more efficiently and cheaply than could whole crystals. Furthermore, cutting crystals into blanks removed the mystery of just what quality rating an individual crystal should be given, thus guaranteeing fair pricing practices. Neither the Signal Corps or the War Production Board looked favorably upon these suggestions, however. Of primary concern was the accuracy with which the crystals could be oriented for cutting. It was felt, particularly by the members of the Quartz Crystal Section, that the quality of such blanks cut in Brazil would be suspect at best, potentially increasing the wastage of raw quartz. Furthermore, with the critical shortage of orienting and sawing equipment, neither agency could condone the export of such material to Brazil when it was needed greatly by domestic plants.

Neither Byington & Co. nor American Gem & Pearl allowed the opposition of the Signal Corps or WPB to deter them. Though Byington was informed in early August of 1942 by Major Olsen of the QCS that their plan for acquiring quartz equipment for shipment to Brazil was opposed, they continued with their efforts.[3] Two months later, Olsen wrote to them again explaining that their latest request for permission to purchase saws for export was opposed by both the Signal Corps and the WPB.[4] Byington had also been active in Rio attempting to set up that end of his proposed operation. In the letter O'Connell wrote to Messer later that October describing the efforts of the QCS over the previous year and alerting him to the imminent arrival of the Signal Corps inspectors, he devoted a page and a half to a discussion of Byington's efforts. O'Connell felt that at heart, all of these proposed Brazilian operations were "get-rich-quick" schemes that were serving no other purpose than to muddle the situation in Rio and contribute to a decrease in quartz exports.[5]

O'Connell also felt that if the Signal Corps allowed companies to cut quartz in Brazil for export to the United States, the QCS would have to accept responsibility for the quality of the blanks. With the problems they were currently having with the raw quartz situation ("We are pretty helpless being this far away, and can't even really form any clear idea of what in hell all the trouble is about"), he certainly did not want to get involved with oversight and inspection of Brazilian cutting plants. Perhaps the best reason of all not to allow such operations was the fact that no crystal companies that he knew of had any interest whatsoever in purchasing blanks from Rio; they all preferred, due to quality control issues, to either cut their own or get them from nearby companies with which they were very familiar.[6] Six months later, Byington was still attempting to acquire equipment for its Brazilian plant.[7]

American Gem & Pearl was just as persistent, though they took a slightly different route on the road to Rio. John Hackes, president of the company, decided to first build a cutting plant in the United States, develop his operation, and then ship the entire concern to Brazil. After securing some contracts for quartz blanks, Hackes was able to acquire the needed saws and X-ray equipment and set up the Knudson Quartz Laboratory in Chicago. Later investigation suggested that some of Hackes' equipment procurement efforts were somewhat outside of approved channels, though not serious enough for legal action to be carried out.[8] After several months of operation in Chicago, Hackes applied to the WPB for export licenses to ship his plant to Rio. His request resulted in a meeting between the WPB, the Signal Corps, and the Board of Economic Warfare at which an official policy regarding this issue was produced.

The policy stated that "no equipment for the cutting of quartz oscillator plates in Brazil should be permitted to be shipped." The rationale for this policy was laid out as a series of ten points:

1. The creation of a quartz-cutting industry in Brazil could have drastic repercussions for the American quartz import program. It was within the realm of possibility that the Brazilian government might limit or restrict the export of raw quartz in an effort to support its own domestic cutting program. This very situation had arisen in Argentina after a beryllium refining plant had been constructed there by U.S. entrepreneurs.

2. Just as political pressure had influenced U.S. purchasing decisions regarding raw quartz (i.e., requiring the purchasing of very low-quality material), similar pressure might be brought to bear for the purchase of cut blanks.

3. The establishment of private cutting plants in Brazil at this time might adversely affect the negotiations currently going on between the U.S. and Brazilian governments regarding quartz purchasing.

4. The Signal Corps was currently working on plans for a pilot plant of its own to support the Brazilian military. Privately operated plants might harm this effort.

5. The transportation of raw quartz (an argument for importing blanks rather than raw crystals) was being handled quite well by the Air Transport Command and merchant shipping. (The Signal Corps had been willing to further discuss the issue of importing blanks if the transport of raw crystals had really been a problem.)

6. The smuggling of quartz to the Axis countries would be much easier if the quartz were cut into small blanks first.

7. The U.S. government was not interested in stockpiling quartz blanks (another suggestion made in favor of Brazilian cutting operations).

8. U.S. manufacturers were not interested in purchasing blanks from unknown plants.

9. Cutting facilities in the United States were adequate to meet the needs of the industry.

10. The difficulties with supply, maintenance, and training related to establishing and supporting such Brazilian operations would be enormous.

One final point that more than anything else sealed the fate of the American Gem & Pearl request was that the products coming out of the Knudson plant were of notoriously poor quality. If anyone were going to be given permission to set up a plant in Brazil, it definitely would not be this company.[9]

Just as the U.S. government had feared, the Brazilian government did become involved in the dispute. Several Brazilian companies had also attempted to gain import licenses for the equipment needed to set up cutting operations; some even sent samples of their work to the QCS for examination, hoping to gain their support. Without exception, however, all such requests for equipment were denied.[10] The Brazilian government was very much in favor of creating a domestic quartz industry and attempted to force the issue in mid-1944. Ironically, a fair amount of the blame for involving the Brazilian government might actually rest with a member of the U.S. Signal Corps, Lieutenant Colonel Herbert Messer.

In January 1943, while still stationed in Rio as the military attaché, Messer took part in a meeting between representatives of the Board of Economic

Warfare and the Brazilian government official responsible for oversight of the quartz program, a Dr. Amaral. At the meeting, Messer brought up the idea of exporting cut bars and blanks. He likened the quartz situation to that of forest products: the value added to an exported log eventually used to make furniture could only be estimated; if the furniture were manufactured in Brazil and then exported, the exact value of the product could be determined (and taxed accordingly). Messer felt that an industry for the production of crystal oscillators should be set up in Brazil after the war, perhaps sooner if possible. Dr. Amaral, as might be expected, was very much in favor of the idea. As O'Connell had written him three months earlier laying out the Signal Corps' opposition to such an idea, it is very hard to understand why Messer would have made this suggestion. Robert McCormick, in his history of the WPB's quartz activities, suggests it was Messer's nearly complete ignorance of the processes involved in the manufacture of quartz oscillators that prevented him from truly understanding the situation. On the other hand, perhaps it was his years of experience in Brazil and his understanding of the local business environment that led him to believe such a venture could be carried out successfully.[11]

Whatever his reasoning, Messer's suggestion took root within the Brazilian government and culminated in a July 1944 resolution by President Vargas calling for the establishment of a domestic crystal oscillator industry. The resolution called for price breaks on quartz sold to domestic manufacturers along with restrictions on exported quartz. The primary restriction was that any country importing Brazilian quartz would be required to also purchase Brazilian-manufactured crystal oscillators at a rate of 100,000 units per month. This restriction was more of a bargaining chip than anything else. Two weeks after the announcement of the resolution, word was sent to the U.S. embassy that Brazil would suspend the 100,000 unit requirement if the United States would grant the proper licenses for the export of sawing and X-ray equipment. The American response was to point out that Brazil could require the purchase of as many oscillators as it wanted; the simple fact that Brazil could not possibly produce that many oscillators made the point moot. Acknowledging the truth behind this statement, the Brazilian government subsequently suspended the resolution, effectively closing the door on all efforts to develop a crystal manufacturing industry in Brazil.[12]

"ALL PREPARATION MUST BE MADE NOW . . . FOR USING SCRAP, SMALL CRYSTALS, AND LOW-QUALITY CRYSTALS PREVIOUSLY NOT CONSIDERED WORTH CUTTING"

The battle for quartz conservation was fought on many fronts. One involved the above-described efforts of the governmental agencies and the Signal Corps to maintain the integrity of the crystal industry by eventually prohibiting inexperienced companies from acquiring quartz contracts and by preventing companies, both foreign and domestic, from establishing crystal plants in Brazil. Another front involved the attempt to convince crystal manufacturers to utilize small crystals of raw quartz. In this attempt, the government was aided by several allies from within the industry itself (one of which was American Gem & Pearl, one of the "enemies" on the other battle front).

The pre-war crystal industry was built around large crystals (two pounds or more). As discussed earlier, the abundance of such large crystals became very low very quickly after the start of the war. Nevertheless, some of the more established companies continued to demand such crystals for their plants. By March 1943, it was apparent to both the Signal Corps and most of the industry that a crisis was looming with respect to the raw quartz supply. At a meeting of an Industry Advisory Committee, a call was made "for stretching the present stocks and new supply of raw quartz and for using scrap, small crystals, and low-quality crystals previously not considered worth cutting."[13] Since the beginning of the war, the Signal Corps had been trying to learn all it could about the influence of crystal size on the speed and efficiency of oscillator production.[14] Early in 1942, batches of small (100–200 grams) crystals were sent to manufacturers for testing.[15] Results from these tests suggested that the number of plates per pound cut from small crystals was essentially the same as that for larger crystals.[16]

Several crystal manufacturers (e.g., Bliley, A.E. Miller, Standard Piezo) had been utilizing smaller crystals since before the war. C.B. Hamilton had used the success of these companies as supporting evidence as he pressed to include smaller crystals in the pre-war purchasing specifications.[17] Word of the ability to utilize smaller stones spread slowly throughout the industry, helped along by QCS representatives and through the subcontracting networks established by companies such as Galvin. One of the strongest advocates of small crystals, however, was Morris Hanauer, owner of American Gem & Pearl. Hanauer truly believed that small crystals could be used if plant owners would just convert their saws and orienting equipment to handle them.[18] One crystal manufacturer that he continually made this pitch to was Louis Patla of DX Crystal in Chicago.

When visiting Patla's plant, Hanauer would urge him to consider using the smaller quartz stones. Patla regularly declined as it would require alterations to his equipment that he did not think were worth the time or the effort. However, as the larger crystals became more and more scarce, Patla began to pay more attention to Hanauer's pleas. On one particular visit, Hanauer convinced Patla to take a coffee break with him and listen to his arguments. According to Patla, "Morris got to talking about this small quartz. He kept insisting he believed it could be used if just someone would try it. He said it was in plentiful supply." Patla was "quite reluctant," but Hanauer persisted. He finally won Patla over with the following plea: "Lou, God made lots of small crystals, but very few large crystals, so why don't you go along with God and use the small crystals?" Patla promised that when things got caught up at his plant, he would convert one saw and one piece of orienting equipment

rtz was distributed to industry test plants over the spring of 1943
ults were just as positive as the earlier ones.[26] The results of these
dramatic effect on the size of the quartz stockpile. An April 1943
the amount of quartz that would be added to the usable stockpile
ast 650,000 pounds." Furthermore, the amount of usable quartz
or purchase in Brazil might increase by as much as 150%.[27]
sible that the usability of such quartz could have been discovered
the attack on Pearl Harbor. As mentioned in Chapter 6, Frederick
ed for just such a research program on the operational effects of
ects in May 1941 and offered the services of the National Bureau
ds to carry it out. The proposed research program was approved
s later by the Office of Production Management with a budget of
r year. Surprisingly, though work supposedly began right away, no
re submitted by the program for well over two years (the first being
January 1944). The early work of the program concentrated on
X-rays in orienting raw quartz, including a detailed study of the
f commercially available X-ray goniometers (instruments used for
angles). By June 1943, however, the lab "was in a position to
re quartz oscillators accurately to any reasonable specifications."[28]
ogress reports were being received by the MRC or the WPB.
ber, R.J. Lund visited the lab accompanied by Robert McCormick
WPB quartz experts. In Lund's opinion, the equipment found at
"far beyond the amount needed to accomplish the results." Among
als being used in the lab were five (apparently new) diamond saws
avy "cast steel, milled tables." Ordered ostensibly for the mounting
, Lund felt that such a purpose was completely illogical (the tables,
sed in the mounting of precision optical devices, would have their
ined in such an arrangement). Overall, Lund felt that the amount
spent in outfitting the laboratory was probably well in excess of the
dgeted by the MRC. A subsequent investigation by the MRC found
be the case, however.[29]
more, the director of the NBS, Dr. Lyman Briggs, felt that Lund's
n of the laboratory's furnishings were incorrect, claiming the lab to
one "precision diamond saw" (and that none of the saws were new)
g that the tables were not milled, nor were they as heavy as had
ned. Robert McCormick took exception to these claims. Writing to
n February 1944, McCormick claimed to have seen the saws himself
ugh only three of them might be correctly described as "diamond
five would definitely be considered "precision" within the crystal
Furthermore, he knew for a fact that at least one was brand new
e was present at the meeting where Paul Galvin assured Frederick
he would have the Felker Company divert one to the NBS lab from
mber he had on order. The issue of the tables being milled or not
y "quibbling over words" as far as McCormick was concerned. As
had granted the purchasing priorities for these tables, he was quite

to work with the smaller stones. After doing so,
the results and slowly began to convert his entii
smaller crystals. By the time that large crystals t
buy, DX Crystals was able to make a seamless tra
ones.[19]

In addition to small crystals, the Signal Corp
ability of manufacturers to utilize the irregula
from Brazil in greater and greater quantities.
more of a problem than small crystals due to the
methods of orientation. As unfaced quartz coulc
had X-ray equipment, these companies were st
Corps to make unfaced crystals as large a perc
possible. As a further incentive, discounts of as
to small, unfaced crystals.[20] By mid-1943, enou;
accept small and unfaced crystals to account f
month.[21] Furthermore, efforts were made to redi
rently possessed by companies unable to utilize
could.[22]

"MR. BLASIER CLAIMED THAT OPTICA ELECTRICALLY TWINNED BLANKS CAI SUCCESSFULLY FOR OSCILLATORS"

Another potential source of raw quartz was the
inspectors due to the presence of various defect:
unfaced crystals, investigations of its usability v
governmental, and civilian participants in the c
As early as April 1941, C.B. Hamilton was
"rejected" quartz to determine just how seriousl
the operation of crystal oscillators.[23] With the c
Corps became involved in such studies as well. I
members, the Signal Corps oversaw several t
rejected quartz. At the urging of Herbert Blasic
look into the possibility of using electrically an
the production of oscillators in late 1942. Blasier
quartz quite often in his pre-war business with
radio customers.[24]

In January 1943, the WPB began authorizin
dard quartz for testing purposes. The first tests
fied as "OB" type quartz: quartz having at l
containing defects such as bubbles and needles
"uniformly good for manufacture of the less e
positive result, further tests were run on the e
quartz (crystals containing less than 10% usabl

of thi
and tl
tests I
estim;
was "
availa

It i
even b
Bates
quartz
of Star
three
$25,00
report;
receive
the use
accura
measui
manuf;
Still, n

In C
and otl
the lab
the ma
and two
of the s
normal
surface
of mon
amount
this not

Furtl
descript
have on
and stal
been cla
the MR
and alth
saws," a
industry
because
Bates th;
a large r
was simi
the WPI

aware of their size, composition, and construction and he did not believe they were suited to the type of use that had been claimed.[30]

In the long run, it was not really the types of equipment that had been purchased that was the issue; it was how they were being used and what results were being obtained that the MRC and the WPB were interested in. Three progress reports were ultimately issued by Bates over the first six months of 1944. Though the reports did cover many aspects of the production of crystal oscillators and the impact of the crystal size and the defects present in the raw quartz, none of the results were new. In fact, most of the results of the NBS study had been known to the industry for the previous two years, having been discovered in the crystal plants and in the Signal Corps Laboratories.[31] For all practical purposes, the research program had been a failure. On July 27, 1944, Lund wrote to S.D. Strauss, vice-president of the MRC, concerning the NBS program. In his letter, he stated the following conclusion:

It was recommended that this laboratory be set up in a letter of May 29, 1941, for the purpose of studying various problems connected with the quartz crystal stockpile. To the best of my knowledge, no real contribution has ever been made and most of the problems encountered have been solved with the aid of the industry. Under these circumstances, we do not feel that we are justified in asking the Metals Reserve Company to support this activity at the Bureau of Standards.[32]

The MRC ceased funding for the laboratory the following month. The laboratory was allowed to keep the MRC-purchased equipment, however, and some work continued under NBS funding.[33] The discussion of this program contained in the wartime history of the National Bureau of Standards suggests a very successful operation. For two-and-a-half small-type pages, the results of the research are described in glowing terms, stressing their importance to the quartz oscillator industry. No mention is made of the delays in reaching the conclusions, the difficult relations with the WPB and MRC, nor of the abrupt cancellation of funding by the MRC.[34]

Nevertheless, the analyses of Lund and McCormick were correct: the program was a failure. Funding in excess of $35,000 was given to the laboratory, yet Bates and his program provided nothing in the way of assistance to the industry.[35] At best, their work can be seen as a controlled, scientific confirmation of the knowledge and techniques that had been discovered and developed within the crystal industry during the first two years of the war. This, however, was never the intention of the program. A golden opportunity was lost in not getting the research program into operation much sooner. Furthermore, as X-ray orientation was already an established component of crystal production (at least on the part of the larger companies), it is hard to understand why this area of research was given a higher priority than that of the crystal defects. If the relatively benign nature of quartz defects with respect to the operation of most crystal oscillators had been determined even a year before it actually was, a great deal of pressure would have been taken off of the raw quartz procurement program. This could have allowed a greater exertion of effort toward increasing the quality of crystal oscillators along with their quantity. Greater control and attention to the quality of product being manufactured might possibly have permitted a much earlier discovery of the aging problem that nearly crippled the industry during the winter of 1943–1944 (and is the topic of the following two chapters).

Such questions can only be speculated about. What can be stated more concretely, however, is that in every case where they had an opportunity to make a real contribution early in the war, the NBS and Bates, in particular, failed to measure up. Bates did claim that delays in receiving equipment slowed his program's progress. This sounds very much like his explanations for the slow pace of the raw quartz inspection program. While it is true that the acquisition of equipment during these years was not an easy process, it is hard to understand how two programs having the complete support of the very agencies that determined who got what and how quickly they could get it could have had so much trouble getting operational. On the other hand, at least with respect to the quartz research program, it appears that the oversight of the MRC and WPB were somewhat lacking. It seems unusual that Lund's first inspection visit to the lab did not occur until October 1943, more than two years after the program was initiated. Perhaps too much of their attention was being claimed by the other aspects of the quartz supply program.

"THERE ARE A NUMBER OF AMERICAN LOCALITIES THAT PRODUCE SINGLE QUARTZ CRYSTALS SUITABLE FOR RADIO OR OPTICAL WORK, NOTABLY THE REGION AROUND HOT SPRINGS, ARKANSAS"

The third front of the battle to provide an adequate supply of raw quartz for the crystal industry involved the search for domestic quartz deposits suitable for producing oscillators. This effort also involved a close cooperation between the Signal Corps and the War Production Board, along with the U.S. Geological Survey. With the emphasis of the 1939 Strategic Materials Act, the Signal Corps initiated a survey to determine whether sufficient domestic sources of quartz existed to preclude its designation as a strategic material. In an attempt to get a "professional" opinion, Colonel Roger Colton wrote the editor of the journal *The Mineralogist* in January 1940. Dr. H.C. Dake responded with what seemed like the answer to the Signal Corps' prayers. It was his professional opinion that, not only should the American quartz deposits be of a similar quality to those of Brazil, but that they should exist in large quantities. "There are a number of American localities that produce single quartz crystals suitable for radio or optical work, notably the region around Hot Springs, Arkansas." Furthermore, he believed that this region could provide "ample" quantities of radio-grade crystals. In his opinion, the only reason that the

American quartz deposits had not been exploited was simple economics: the Brazilians charged much less for their crystals.[36]

Dake's letter was forwarded to the OCSigO, where it was received with a fair amount of surprise. Why, if Dake's opinions were justified, had no government studies ever come to the same conclusions? Colton was requested to follow up on the letter and see if Dake might be able to supply the Signal Corps with samples upon which they could perform tests. Dake's letter was then forwarded to Lyman Briggs at the NBS, requesting his opinion of the letter's claims. Briggs responded that the NBS had been involved for years in a search for domestic deposits of quartz suitable for optical work (having purity requirements only slightly higher than that used for radio purposes). In his opinion, though there must surely be "large quantities of this material" somewhere within North America, the "best efforts" of the NBS had failed to find them. Briggs echoed the suggestion of asking Dake to supply samples of the domestic quartz he claimed to exist.[37]

The Navy Bureau of Engineering was also asked for its opinion on the Dake letter. Their response was that, although Dake was correct that quartz was a very common mineral in the United States, very little of it had ever been found that satisfied the specifications for crystal oscillators. It was either "too small for practical use," lacked faces, or possessed twinning. As far as they were concerned, the only dependable source of radio-grade quartz was Brazil.[38] Thus, armed with this advice, the Signal Corps retained quartz as a strategic material.[39]

The abrupt start of the war and the accompanying fears regarding the sufficiency of the Brazilian supply led to a re-evaluation of the situation. Just how much radio-grade quartz had been found in the country? Should Dake's suggestion concerning the Hot Springs area be followed up on? Perhaps all previous searches for "suitable" quartz had failed because the standards had been set too high? With the rapidly changing specifications determining just what types of quartz were acceptable, might a new nationwide search be worthwhile?

The first promising lead came while Signal Corps Lieutenant J.J. Lloyd was making a tour of the west coast in April 1942. While in San Francisco, Lloyd heard about an abandoned gold mine in Calaveras County that was "rumored" to have also produced radio-quality quartz. Lloyd passed the word along to James Bell at the WPB who, along with Robert McCormick, looked into the situation.[40] As it turns out, the Navy had purchased approximately 1,000 pounds of quartz from this mine in 1926 and 1927; 400 pounds of which had actually been used for oscillators. With the approval of the U.S. Geological Survey, loans totaling $25,000 were made to the mine owners in an attempt to bring it back into operation. Unfortunately, the mine never produced enough usable quartz to keep the government interested in backing the operation. Though the owners attempted to carry on the effort using their own capital, they soon gave up as well. Nothing close to the yields from the 1920s was ever achieved.[41]

As they traveled the country during the early months of the war, attempting to build a crystal oscillator industry, the men of the QCS kept their eyes and ears open for the possibility of domestic quartz deposits. One promising area in the Northeast was the region around Little Falls, New York. While in the area in April 1942, Lloyd and Willie Doxey made contact with a local amateur crystallographer, Donal Hurley. Hurley was quite knowledgeable about North American crystal deposits, and was particularly interested in quartz. Hurley took the QCS men on a tour of the area and helped them collect samples to be tested at the Signal Corps Labs.[42] He also supplied the QCS office with names and addresses of persons knowledgeable about deposits in Michigan and Massachusetts. He continued to send in samples of quartz throughout the summer. Unfortunately, subsequent tests showed that all of the samples from New York were unsuitable for crystal oscillators.[43]

By far the most promising region of the country in which to prospect for quartz was Arkansas, particularly Garland and Montgomery counties. Quartz had always been a fact of life for the inhabitants of this region. Young boys growing up in the 1920s would quite often cut their bare feet on crystals unearthed while plowing their families' fields.[44] Even today, Montgomery County claims the title of "Quartz Crystal Capital of the World."[45] In November 1942, McCormick and a team from the Geological Survey traveled to Hot Springs to see just how much of this quartz might be available. Based on their initial surveys and the reports of local quartz prospectors and dealers, the WPB encouraged the MRC to initiate a mining operation in the region as soon as possible.[46]

The project was approached with a great deal of excitement. As quickly as possible, samples of the Arkansas quartz were sent to crystal plants to be manufactured into oscillators. The results of these test runs were eagerly anticipated by both the WPB and the Signal Corps.[47] From all indications, the Arkansas quartz appeared to be of similar quality to that of Brazil and produced similar numbers of oscillator plates per pound.[48] In fact, a pictorial in the August 2, 1943, issue of *Life* magazine, which stated that "the U.S. has only poor quality quartz" resulted in a letter to the editor from McCormick. McCormick's letter appeared in the August 23rd issue and stated, in part, that "the quantity of quartz crystal being produced in the U.S., while small, is steadily increasing, and the quality has been found to be equal in every respect to Brazilian crystal."[49] Nonetheless, if the government were to continue financing the Arkansas mines, the quantity of this quartz would be the determining factor. By the end of September, 1943, $150,000 had been spent on the operation. Unfortunately, only a few hundred pounds of high-grade quartz had been extracted, much too small an amount to justify continued operations. John Faith, an MRC representative, was left in Hot Springs through May of the following year as a purchasing agent. It was hoped that a continued governmental presence in the area might stimulate private prospectors and possibly uncover other potential deposits.[50]

Hope of locating other domestic deposits was kept alive through most of the war. Advertisements were placed in newspapers around the country and in mining and geology trade journals asking for information on quartz deposits.[51] Responses came in from all across the country (e.g., Virginia, the Carolinas, Wyoming, Pennsylvania, Nevada), many with accompanying samples of crystal.[52] The results were always the same, however; the samples were never of sufficient quality for the production of oscillators. By July 1944, the Signal Corps believed it was time to pull the plug on any future prospecting. In a letter to the Small War Plants Corporation, their position was stated that "as of today, this office believes that locating deposits from which radio quality crystal has not previously been extracted is not of value to the war effort." The OCSigO would no longer endorse the expenditure of funds for such ventures.[53]

Along with searches for quartz deposits within the United States, investigations were carried out throughout the western hemisphere in hopes of discovering additional sources. Mexico, Trinidad, Newfoundland, and Cuba were sites of possible deposits. As with the domestic deposits, none ever amounted to anything in terms of quantity or quality.[54] An episode took place over the final months of 1943 that raised hopes that a new source of quartz might have been discovered. In October, word reached the QCS that a 58-pound sample of quartz, believed to have originated in the African Gold Coast, had been received at Wright Field. Swinnerton wrote the Aircraft Radio Laboratory there asking for information regarding the quality of the quartz and its place of origin; both the OCSigO and the Board of Economic Warfare were interested as well.[55] The quartz appeared to be of high quality and had been shipped to the NBS for further inspection. Unfortunately, no one at the ARL knew where it had come from. Colton wrote to the commanding officer of the Eighth Air Depot in Miami, Florida, the shipment's point of entry into the United States, asking for any information he might have.[56] Ultimately, all hopes of an African supply of high-grade quartz were dashed when it was discovered that the shipment had originated from Brazil.[57] It is not clear from the records how the initial conclusion that the quartz came from the Gold Coast was reached. However, it seemed valid enough to keep the Signal Corps and the BEW buzzing with activity for three months.

"NOTHING IN THE PRESENT SITUATION JUSTIFIES THE EXPENDITURE OF EFFORT AND MONEY TO PRODUCE ARTIFICIALLY GROWN QUARTZ CRYSTALS"

During the early months of the war, serious consideration was given to the possibility of utilizing "quartz substitutes," other piezoelectric minerals that might be used for frequency control. Though research programs were carried out, none of the minerals offered any potential for large-scale use. In O'Connell's opinion, the "choice of a substitute is therefore restricted to [quartz] crystals to be grown artificially."[58] Work had recently been carried out on the subject of "synthetic quartz" by Paul Kerr of Columbia University (such material, grown from a seed of natural quartz, would today be referred to as "cultured quartz"). The Navy found Kerr's work of interest and passed a report on it along to Bates at the NBS for his opinion. Though Bates found it interesting from a purely scientific point of view, he saw no cause to pursue it during wartime. Finding that "nothing in the present situation justifies the expenditure of effort and money to produce artificially grown quartz crystals of a size suitable for the production of electric oscillators," Bates recommended that no such program be encouraged.[59]

Ten months later, the idea resurfaced. Dr. Kerr was asked by the QCS to supply a copy of his research report on quartz synthesis and McCormick contacted scientists at the Geophysical Laboratory to gauge their willingness to carry out a government-funded study. Though the scientists were willing, the administration of the laboratory felt that their time could be better spent on other projects. It seems as though Bates' opinion from the previous year carried a great deal of weight within the government's scientific establishment. The reasons given by the administration of the Geophysical Laboratory, and later by the board of the National Defense Research Committee, for not approving the project echoed very closely Bates' May 1942 letter to the Navy.[60] Though a synthetic quartz program was not initiated by the United States during the war, one was carried out by Germany, to some success (as will be discussed in the final chapter).

In a darkly humorous episode that demonstrates the lengths some people went to in order to take advantage of the wartime boom in funding for industrial and scientific research, two men did claim to be able to synthesize quartz in any quantity needed by the government; all they needed was $50,000 to get their project underway. In March 1944, a letter was received by the WPB from E. Van Dornick, a chemical engineer from New York City. In the letter, Van Dornick claimed to have invented a process for producing artificial quartz and had been referred to the WPB by Clifford Frondel (who by this time was employed by Reeves Sound Laboratories in New York City). According to Van Dornick, the only limitation on the size of the crystals he could produce was his equipment. If he could only receive the proper blanket priorities and financing from the government, the quartz supply problem would be solved forever.[61]

In response to this letter, a meeting was called in order for representatives from the WPB and the Geophysical Laboratory to meet with Van Dornick and his partner, J.W. Herrick. At the meeting, Van Dornick produced a nearly perfect quartz crystal of approximately 10 cm length. When questioned as to their production process, the two partners claimed they were unwilling to divulge the details just yet as they had no patent protection. However, they still urged the WPB to award them the requested funding based on the sample crystal they displayed. Ultimately, it became clear to them that they were not going to receive a check and thus excused themselves. They were convinced to leave their crystal behind for further examination, however. What the examination showed was that the crystal had formed naturally, but had been altered by cutting and grinding to give it a more nearly perfect and "artificial" appearance.[62]

Van Dornick and Herrick were informed of these results and advised that the government would no longer consider any backing of their project. This did not stop them from pursuing private funding, however. Reeves Sound Labs was one of their primary targets. A few weeks after the Washington meeting, Frondel wrote McCormick alerting him to the duo's activities. On their first meeting with Frondel, the pair had produced what they claimed to be an artificially grown crystal. Examining it under a microscope, Frondel was immediately able to determine that it was natural *and* from what region of the country it had most likely originated. Though he did not make this known to them, he did ask them to bring him another crystal, this one grown around a string. He further advised them that the WPB was in charge of all quartz-related programs and that they should contact McCormick in Washington.[63]

After their meeting with the WPB, Van Dornick and Herrick returned to Reeves and told Frondel that their quartz had been verified by the Geophysical Laboratory and that McCormick had proclaimed in the meeting that their process worked. Based on this information, the partners believed that Reeves should finance their project. When reminded about his request for a crystal grown around a string, Van Dornick became upset, stated that he would provide no such thing, and that the endorsement of the WPB and the Geophysical Laboratory should be good enough for anybody. It wasn't for Frondel and they got no financial support from Reeves Sound Labs.[64]

They did move ahead with applications for priorities to acquire their needed equipment. Before their application was acted upon by the WPB, the Office of Production Research and Development received a phone call from "a prominent Midwestern oil company" stating that Van Dornick and Herrick had approached them for financing, claiming to have already been granted the proper priorities from the WPB. In quick succession, their chances of getting any money out of this company and of getting any approval from the WPB were reduced to zero. No further word was ever heard from the "engineers." Though the file on this episode was provided to the FBI, no actual evidence of fraud could be obtained and no charges were ever filed.[65]

Perhaps, with the proper backing, and an effort on par with that exerted by Van Dornick and Herrick, a synthetic crystal program might have been developed during the war. In hindsight, however, the decisions made regarding this line of research were probably correct. The efforts at conserving raw quartz throughout the manufacturing process and the tremendous increase in the stockpile due to the discovered usability of most of the "rejected" quartz guaranteed a suitable supply of quartz throughout the duration of the war. The importance of these efforts cannot be overemphasized. If the industry had continued to operate under the size and quality specifications in effect at the beginning of the war, the quartz crystal program would have been a disaster. The steady decrease in allowable crystal size, led by the efforts of government men such as Hamilton, private companies such as Bliley and A. E. Miller, and importers such as Hanauer of American Gem & Pearl, meant an ever-increasing supply of raw material.

The determination that optically twinned and defect-bearing crystals could be manufactured into oscillators, resulting primarily from the studies overseen by the Signal Corps' Quartz Crystal Section, added more raw quartz to the usable stockpile than could ever have been supplied by the domestic sources searched for with such dedication. The willingness to try new techniques, adapt to lower-quality crystals, and to communicate vital information among its various entities are some of the characteristics that resulted in the quartz crystal industry producing nearly two million oscillator units per month by the fall of 1943. Unfortunately, due to a less-than-complete understanding of the physics of quartz oscillators, nearly all of the units coming out of the crystal plants at this time were doomed to failure.

9

THE AGING CRISIS—STOPGAP MEASURES

The first reports of crystal failures came in from the Pacific theater. Then from the China-Burma-India (CBI) theater, Great Britain, and, ultimately, from units stationed in the United States. Throughout the summer and fall of 1943, communications failures due to crystal malfunctions steadily increased. By this time, the crystal problem was supposed to have been solved. A mass production industry had been built and was being supplied with raw quartz at a rate enabling it to produce over a million oscillator units per month. Field units were finally beginning to receive their full complement of crystals and significant stockpiles were being built up at Signal Corps depots across the nation. This sudden and unexpected outbreak of failures must have struck the men of the OCSigO as a cruel trick of fate.

"THE EXTREMELY HIGH RATE OF DETERIORATION SUFFERED BY CRYSTALS DUE TO AGING HAS BEEN AND IS A SOURCE OF MUCH TROUBLE TO THE ARMY AIR FORCES"

The early reports, coming from regions of extreme humidity, suggested moisture-related problems. A portion of the failures were obviously due to this effect; brass contacts were found to have corroded under the effects of moisture and the chemicals of which the holders were composed. Changes in the metal used for the contacts and in the composition of the holders solved this particular problem.[1] The vast majority of the failures, however, were cases where the crystal units had simply gone "dead"; they would no longer vibrate when connected within the electrical circuit of the radio. Though it was found that these crystals could be caused to vibrate again by washing their surfaces, their oscillating frequencies were always found to have increased.[2] This phenomenon was known as "aging" and represented perhaps the most serious crisis faced by the crystal industry. All of the efforts of the previous years to build an industry and supply it with raw materials would have been wasted if the products of that industry were destined to fail.

The phenomenon of aging, a decrease in activity accompanied by an increase in oscillating frequency, had been known for as long as crystal units had been used. However, nothing like the current cases had ever been experienced. Two primary factors accounted for the increased seriousness of the problem. First, the rate of aging depended on the frequency of the crystal with higher frequency units aging faster than those of lower frequency. Radios were now operating at frequencies much higher than ever before, especially those utilized by the Army Air Forces. Second, transmission channels were being crowded closer together within the radio spectrum than ever before. Thus, a change in a crystal's frequency that, years before, might not have caused a problem now moved the oscillator unit completely out of its assigned channel. The very characteristics that had led to the broad acceptance of crystal control, high frequency use and close channel packing, were now contributing to its failure.

Efforts were initiated to determine the extent of the problem. While reports were requested from field units overseas, teams of inspectors were organized to survey the holdings of the various Signal Corps supply depots.[3] The survey results confirmed everyone's worst fears: the problem was severe and it was widespread (both oversees and within the depots).[4] Most seriously affected were the units of the Air Forces; of the two arms most heavily dependent upon crystal control, the Air and the Armored Forces, the Air Forces utilized much higher frequencies and required much tighter frequency tolerances for their channel assignments.[5]

By early 1944, the situation in Great Britain had reached such an extreme state (regular failure rates of shipments ranged from 10 to 40%, with 80% failure rates having been experienced) that the Air Forces requested the establishment of a facility dedicated to the testing and repair of aged crystal units. In a memo from the headquarters of the Army Air Forces (stamped "IMMEDIATE ACTION"), Chief Signal Officer Ingels was informed of the "extremely high rate of deterioration suffered by DC-11, DC-16, and DC-26" crystal units (the units used in the SCR-522 airborne command sets, the SCR-542 jeep-mounted air liaison sets, and the SCR-624 transmitters used by air traffic controllers). The Royal Air Force was experiencing similar problems and had requested a set of "reactivation" equipment for its own use. Along with the repair facility, the AAF headquarters requested that a crystal engineer (civilian if necessary) be sent to England to oversee the construction of the facility and the repair efforts of its personnel.[6]

"DIP THE TOOTHBRUSH IN THE SOAPY SOLUTION AND BRUSH EVENLY BACK AND FORTH ACROSS THE SURFACE OF THE CRYSTAL"

As mentioned above, the aging phenomenon had been known about for some time, and manufacturers had always been interested in finding ways to prevent it from happening to their products. During the early years of the war, four different remedies for the effect were utilized: improved grinding techniques, wiping, scrubbing, and etching (some manufacturers even utilized a fifth technique of simply leaving the frequency of the plate a few hundred cycles too low, allowing the aging process to bring them up to the proper frequency).[7] During the grinding process, it was believed, "pinnacles" of quartz could be left sticking up from the surface due to the use of abrasives that were too coarse. Should these pinnacles later break loose from the surface, an increase in the frequency (due to the decreased mass) and a decrease in the activity (due to the damping effect of the material now covering the plate surface) of the oscillator could occur. Companies were experimenting with different types of abrasives as early as January 1942 and having apparent success in diminishing the effects of aging.[8]

The wiping technique was actually developed as a means of adjusting quartz plates to final frequency. It was discovered by accident that the wiping of an oscillator plate with a cloth could lower the frequency of the plate slightly (due to the "loading" of the surface with residue from the cloth). The method came to be used for correcting plates that had been ground beyond their intended frequency and actually increased the production rate of the crystal units. The procedure was never sanctioned by the Signal Corps; in fact, Virgil Bottom was once summoned to a Kansas City crystal plant by an FBI investigator who had observed crystal finishers wiping oscillator plates and had suspected sabotage. The women had discovered the effect on their own and believed they were doing nothing wrong with their procedure. Nevertheless, Bottom put a stop to the activity and all crystal plants were warned by the Signal Corps not to load oscillator plates with any foreign material. As it turns out, the technique had spread to several crystal plants in the area before the FBI inspector blew the whistle. Months later, when depot stocks were being tested for changes in frequency due to aging, it was discovered that the units from these Kansas City plants had aged significantly less than those from plants not employing the wiping technique.[9]

The effect of the wiping was to remove some of the loose particles of quartz left on the surface of the plate due to the grinding process. Once this fact was recognized, the Signal Corps reversed its stance on wiping; in fact, it required that the surfaces of quartz plates be *scrubbed* in order to prevent aging.

A September 1943 report entitled "Cleaning of Crystal Oscillator Plates" required Signal Corps inspectors to check oscillator plates for surface cleanliness using a toothbrush and soapy water. If an oscillator plate could be scrubbed vigorously and not experience a significant change in frequency, then its surface was free of both foreign matter and loose pinnacles and therefore should not age. Crystal manufacturers decided that, if the inspectors were going to use such a test, then they would incorporate it as a standard step in their own production processes. Virgil Bottom remembered that, in towns possessing crystal plants, it soon became very difficult for the ordinary citizen needing a toothbrush to find any left in the stores.[10]

The final technique found useful by crystal manufacturers for finishing crystals (and preventing aging) was that of acid etching. As discussed in Chapters 4 and 5, the Bliley Electric Company utilized the etching-to-frequency technique primarily because of how it simplified and increased the speed and accuracy of crystal finishing. Other companies, such as the Harvey-Wells plant of Southbridge, Massachusetts, also used the acid etching technique for the very same reasons.[11] Overall though, the process was not very common among the industry, particularly among those companies that joined the industry after America's entry into the war. For the most part, the grinding and final lapping processes were easier to set up and get into production. This is unfortunate in that, as will be discussed in detail in the following chapter, acid etching was the only reliable method of completely preventing aging.

Another reason that the acid-etching technique was not encouraged by the Signal Corps was that the other methods, particularly the scrubbing, appeared to work. In-plant inspections suggested that the oscillators being produced did not age appreciably. This was due to an unknown factor of the physics of the quartz surfaces. It would be discovered later (and will be discussed in the following chapter) that a quartz plate underwent two phases of aging: a rapid initial phase and then a slower, more steady phase. The wiping and scrubbing techniques only served to eliminate the first phase; the slower second phase still took place. Thus, the Signal Corps found itself in the middle of a very serious and similarly unexpected crisis by the summer of 1943.

"THE MISSION OF THE CRYSTAL GRINDING UNIT IS TO PERFORM ALL ECHELONS OF MAINTENANCE AND TO MANUFACTURE QUARTZ CRYSTALS"

Fortunately, when the crisis hit in mid-1943, the Signal Corps was very close to having put into operation a plan that would serve as a fairly effective stopgap measure until a true understanding of the aging phenomenon could be reached. This plan involved the use of "crystal grinding teams," small units of Signal Corps personnel operating crystal finishing and repair shops within the various theaters of war.

In May 1942, O'Connell wrote the director of the General Development Laboratory at Ft. Monmouth about the possibility of developing quartz finishing units for the close support of front-line troops. The idea was, with crystal control becoming so widespread, it might soon become necessary to offer support related to crystal oscillators even when the troops were far from established supply depots. The laboratory was directed to begin working on the details of just how such a unit would be equipped for producing crystal oscillators in the field.[12] One idea was to build a complete inspection and fabrication laboratory within a large ambulance truck. This idea was later rejected in favor of a set of three smaller trucks specifically outfitted for the finishing of quartz plates and their installation into holders. Repairs of damaged units and the regrinding of plates to new frequencies would also be carried out in this mobile lab.[13]

The actual teams of Signal Corps personnel would be kept small: one officer (a 1st Lieutenant) and three enlisted men. Their mission required them to "perform all echelons of maintenance and to manufacture quartz crystals." The crystal grinding teams would function completely independently of the units to which they were attached. No "re-employment, replacement, or reclassification to other duty" would be allowed without the express consent of the Chief Signal Officer.[14] The officers were appointed directly by the CSigO while the enlisted men were selected from the Signal Corps' replacement training centers and technical schools.[15]

Some men, becoming aware of the program, volunteered for the grinding teams. This included men already in the Army (such as George Fisher) looking for a transfer to better duty and civilians facing imminent induction. As discussed in Chapter 5, men such as Fisher could simply be transferred from one unit to another. Men not yet inducted, however, could not be acquired by a particular unit quite so easily. Such men were informed by the QCS that they should request assignment to the Signal Corps as soon as their basic training was completed. Once a member of the Signal Corps it should be possible for them to be assigned to crystal grinding duty.[16] Men were not the only volunteers for the grinding teams. Two female civilian employees of the Camp Coles laboratory, Selma Greenwald and Vivian Nowicki, inquired as to the possibility of joining the WACs and being assigned to a team. They were informed of the nonfeasibility of this idea, however, by the QCS. Though he expressed appreciation for their suggestion, Major E.W. Johnson, assistant officer in charge, explained to them that there existed very strict regulations regarding the deployment of WAC units and the duties of the crystal grinding teams would not allow them to satisfy these requirements (e.g., there would not be adequate housing for female personnel, nor would there likely be any other WAC units in the vicinity of the locations where grinding teams would most likely be stationed).[17]

The first teams were trained at the Camp Coles Signal Laboratory near Ft. Monmouth. After November 1943, the training program was relocated to the Holabird Signal Depot in Baltimore, Maryland. A training course lasted for eight weeks and included all aspects of oscillator manufacture. Team officers also received tours of local crystal plants as part of their training. As rumors of oscillator failures began to increase, the priority of this program rose dramatically. The Camp Coles program was instructed to have the first four teams trained and equipped for deployment by June, even if they had to issue them pieces of the laboratory's own equipment. The Camp Coles instructors were successful, though they did in fact have to resort to the option of donating some pieces of machinery from their own laboratories. The first four teams shipped out for Australia in June 1943; all were in-country and operational by the end of the year. Five more teams were deployed in July (four to North Africa and one to Great Britain), with another 12 shipping out by the end of the year to such locations as Panama, Hawaii, Alaska, India, and the Middle East.[18]

Deploying a crystal grinding team required no small effort. Over 12,000 pounds of equipment was required for each team's operation. Great care had to be taken in the loading, shipping, and unloading of the equipment (much of it rather delicate in nature). Many teams arrived at their stations only to find a great deal of their equipment damaged due to rough conditions while underway, or more commonly, rough treatment by the dock crews who unloaded it. Where possible, damaged equipment was repaired. In other cases, spares and replacements had to be requested from the United States.[19] Feedback from the teams did result in improved packing materials and methods to be used at the Signal Corps depots and laboratories where the equipment would originate.[20]

"WE FINALLY RESORTED TO VALVE GRINDING COMPOUND FROM THE MOTOR POOL"

By the fall of 1943, reports (most stamped "SECRET" or "CONFIDEN-TIAL") began arriving at the OCSigO detailing the experiences of the crystal grinding teams during their first weeks of operation. For most teams, semi-permanent quarters were arranged and their workshops were set up. Such workshops ranged from buildings at large signal depots (such as in Australia and Alaska) to large tents with wooden floors raised to keep the equipment above the monsoon flood waters (such as in New Caledonia).[21] Others, such as those assigned to George Patton's 3rd Army found themselves unable to carry out their duties after the "breakout" from the Normandy beachheads. The army moved much too fast for them to have set up and utilized their equipment, even if it could have kept up with them. Some groups were reassigned to other repair duties; one grinding team was finally able to establish a working shop by December 1944.[22]

The teams were kept busy supporting the fighting units in their areas, whether working from established depots in rear areas of the Pacific or moving with advancing units across Italy and France.[23] The Australian and Southwest

Pacific teams in particular were kept busy as they served as the sole support units for the U.S. Army, Navy, and Marines along with forces from allied nations.[24] To better serve the needs of the area, the four units assigned to the 832nd Signal Service Company in Australia were combined into a single 12-man unit. As four officers were considered too many to oversee 12 men, three were reassigned to other duty in the area.[25]

The standard duties of the crystal grinding teams included repairing defective oscillators, regrinding crystal units to new frequencies, and producing new units to required specifications. The Australian and Pacific teams in particular dealt with a large number of aging-related failures (up to 80% of failures reported to the 842nd Signal Service Co. stationed on New Caledonia were due to aging).[26] In these cases, new blanks were ground to the required frequency and the aged plates were kept for possible regrinding. Waterproofing was also an important part of the work done by the units in the more humid theaters (including the China-Burma-India theater). Though it did not completely prevent humidity-related failures, waterproofing did extend the life of a crystal unit.[27]

Along with the repair of failed crystal units, a grinding team's ability to manufacture units for special orders made them a very valuable resource for the combat units. At times, due to a shortage of materials or a request for a particularly uncommon frequency, the grinding teams had to use a great deal of ingenuity. One such outfit was assigned to the 842nd under the direction of Sergeant H.J. Benedikter. Benedikter had volunteered for a "secret assignment" after finishing Radio Repair School in the spring of 1943. The secret assignment turned out to be the crystal grinding school at Camp Coles. By November, he found himself assigned to the South Pacific Base Command on New Caledonia. An order came in to his shop one day for a set of oscillators needed by the Air Force to operate at 455 kHz. The only blanks he had near this frequency were some relatively thick ones with frequencies of 200 kHz. A great deal of quartz would have to be ground off these blanks in order to bring their frequencies up to the 455 kHz required.

There were two problems, however. The holders on the lapping machines that held the blanks and moved them around as they were ground were not designed to work with blanks of such a thickness. Furthermore, the grinding compounds available to the team were all too fine to remove the amount of quartz needed in any reasonable amount of time. Nevertheless, the Air Force needed these oscillators and Benedikter's team was expected to produce them. To solve the first problem, the team designed and built a new carrier for the lapping machine from Bakelite (a plastic-like material that hardened upon baking at high temperature). For the second problem, a very coarse grinding compound was located at the motor pool; normally used for smoothing engine cylinders and cannon barrels, it was of sufficient grit to quickly bring the blanks down to the needed thickness. Upon finishing the blanks to the proper frequency, a third problem was discovered: the holders available for mounting the plates were too small for the plates to fit. This problem was solved as well by enlarging the inside of the holders with a machine-shop lathe. The Air Force received its needed units, and on time as well. Benedikter eventually received the Bronze Star for his service on New Caledonia as chief noncommissioned officer of the crystal grinding team.[28]

At times, units found themselves so isolated that even the crystal grinding teams were unable to support them. Burma was just the type of theater where this could happen. Fortunately for John Holmbeck's unit, he was able to step in and serve as an "amateur" crystal grinding team. Holmbeck learned to grind crystals while working on his ham radio license during the late 1930s and put his skills to work for his unit. Using grinding compound and a piece of glass from an old jeep windshield, Holmbeck would salvage crystals from broken radios and regrind them to needed frequencies.[29] Though his efforts were generally successful, such activities were officially forbidden. A great many crystal units sent to grinding teams for repairs showed evidence of previous attempted repairs by the soldiers in the field. In some cases, the attempted repairs had rendered the units completely unsalvageable. Strongly worded orders were sent out to all commands making it very clear that no one was to open or otherwise attempt to repair any crystal units; they should all be sent to the nearest crystal grinding team.[30]

Though the Navy made use of the Army's grinding teams when they could, they also attempted to set up crystal shops of their own. Ken Thomson was a civilian Navy employee who later became a major player in the Kansas City crystal industry. During the war, Thomson was involved in setting up crystal shops, both aboard ships and at Naval bases. His work took him throughout the Pacific; he occasionally found himself attempting to grind crystals while his ship was under attack by Japanese aircraft.[31]

The crystal grinding teams played a very important role in keeping combat units supplied with functioning crystal oscillators. If the aging crisis had never occurred, they would still have been needed for making repairs and manufacturing special orders. However, the aging problem made them absolutely necessary. Units failed at rates much higher than would have been expected due to normal wear and tear. As a stopgap method of keeping the radios functioning, the teams were invaluable. Even so, the Signal Corps could not be content with stopgap measures, no matter how successful. A solution to the aging phenomenon had to be found. To do so would require a more complete understanding of the physical principles at work within the vibrating quartz plates. The ultimate solution to this crisis fell to the scientists and engineers of the Signal Corps laboratories. The fate of the wartime industry (and any future commercial industry) rested with them.

10

THE AGING CRISIS—PHYSICS TO THE RESCUE!

The Signal Corps Laboratories were no strangers to cutting-edge scientific research. A legacy of such activities existed stretching back to the early days of the telegraph. Robert Millikan, an eminent physicist and Nobel Prize winner served as a major in the Signal Corps during World War I. A great deal of research regarding radio transmission was carried out at the Signal Corps Labs during the period between the world wars. Even General Colton himself had published articles in the field.[1] The aging crisis presented the Signal Corps scientists with their most urgent challenge to date. A full understanding of its causes needed to be gained; only then could a cure be developed.

"BY USING A FIXED STEP ABRASIVE GRINDING PROCEDURE, MOST OF THE TROUBLE MAY BE ELIMINATED"

The characteristics of aging had been known for some time. For a group of plates of different frequencies, all ground in the same manner, the highest frequency plates aged quickest. For a group of plates ground to the same frequency using abrasives of differing coarseness, those ground with the coarser abrasives aged faster. Moisture accelerated the aging process, and aged plates were found to have a thin covering of power on their surfaces. Under spectrographic analysis, the powder appeared to be quartz.[2]

Crystal Clear: The Struggle for Reliable Communications Technology in World War II,
by Richard J. Thompson, Jr.
Copyright © 2007 by Institute of Electrical and Electronics Engineers

Formal studies had been going on at the Signal Corps Laboratories at Ft. Monmouth since 1940, though they were primarily concerned with finding methods for eliminating or at least minimizing its effects. Theoretical work had been carried out in the early 1930s that demonstrated that disorganized surfaces resulted in a damping of vibrations within the crystal plate. Such disorganization could result from grinding along with possible recrystalliza-tion of the outer layers of the quartz. Research at General Electric suggested that water absorbed through the surface or through capillary action along cracks resulted in a disruption of the surface layer.[3] Work was done on the effects of heating the plates in hopes of relaxing the surface and decreasing the amount of disorganization. This technique showed little promise.[4]

By late 1940, research began to focus on ways of removing the disorganized layer of "pinnacles" before they had the chance to come loose during opera-tion. Methods of using successively finer abrasives were tested, showing some success. A technique of plating a silver layer on the crystal and then peeling it off to remove the pinnacles was also tried. This showed even less success, plus it was too expensive and time consuming to be used in a production scheme. Work was carried out on acid etching, but was found to be very sensi-tive to the concentration of acid and the length of etching time.[5]

By the end of 1940, it appeared that the most promising method involved the combination of the four-step abrasive and acid etching. A November 1940 report stated "that by using a fixed step abrasive grinding procedure, most of the trouble may be eliminated." However, it was believed that etching still should be incorporated in the manufacturing process as a safety factor. The report did point out, however, that the one-month time period over which the study had so far been carried out was "too little time to draw any definite conclusions" about the overall dependability of the method.[6] By April 1941, however, the details of the etching technique had still not been worked out. A confidential laboratory report stated that "etching will be incorporated in the processing as soon as details can be worked out and operators trained."[7] With the coming of the war, Signal Corps research on aging was greatly diminished in favor of research on mass production methods of oscillator manufacture.

"OWING TO THE CRITICAL SITUATION . . . THIS PROBLEM SHOULD BE GIVEN A HIGH PRIORITY"

By 1943, aging was once again a front-burner problem for the labs. At the urging of Karl Van Dyke, Captain E.W. Johnson, temporarily in charge of the QCS, wrote the Director of Camp Coles Signal Lab in April 1943, instructing him to begin "an extended study of the aging problem as related to the steps in grinding and finishing." A current theory was that tiny pieces of ground quartz powder were getting stuck within cracks in the surface of the quartz and, upon coming loose, would damp out the oscillations of the plate; Van

Dyke wanted an idea of just how valid a theory it was. Johnson asked the laboratory to look into this idea and keep the QCS informed.[8] By September, the problem had grown much larger and the Signal Corps was even more anxious to find a solution. This time, General Colton, writing for the Chief Signal Officer, urged the Camp Coles group to increase its efforts regarding either preventing aging, or at least accelerating its results (so that the final frequency of an oscillator plate would be known before being installed in a holder). "Owing to the critical situation which a failure of current manufacturing and testing methods would introduce," Colton wrote, "this problem should be given a high priority." Progress reports were to be submitted regularly to Virgil Bottom (now on the QCS staff). Though reports did come in, they were of no satisfaction to anyone at the OCSigO.[9]

The aging study at Camp Coles was being carried out by Dr. Dominic D'Eustachio, a physicist trained at the Brooklyn Polytechnic Institute. According to Bottom, D'Eustachio took a very maddening approach to his research: he would claim to understand the problem and be on the verge of a breakthrough, but would refuse to share any information on what he'd learned. Rumor had it that he expected to receive a Nobel Prize for his work and was worried about protecting his interests.[10] In November 1943, he did produce a preliminary report on his findings. According to his theory, the grinding of a quartz blank left a disorganized surface layer upon which new crystal growth occurred. Taking place at the sites of "unsaturated bonds," this crystal growth led to "a large number of very small quasi-randomly oriented crystallites." This unorganized crystal structure, according to D'Eustachio, caused a great deal of pressure to build up under the surface of the plate, ultimately causing material to crack off and remain as a dust deposit on the surface.[11]

D'Eustachio claimed to be able to support his crystal growth model through the use of X-ray photographs. As discussed in Chapter 5, X-ray diffraction is an excellent method for probing the atomic structure of a crystal. If new crystal growth were occurring on the surface of ground quartz plates, successive X-ray photographs should show such changes in the structure. D'Eustachio possessed photographs that he claimed showed just that. In his report, he explained the effects of increased humidity on aging (the water weakened the surface through some unknown process and accelerated the cracking due to the growth-induced stress). He also offered a solution: coat the surface with a substance that would react very quickly with the "unsaturated bonds" on the quartz surface and prevent the growth of new crystals. Furthermore, he believed the 1940 suggestion of plating the surface and peeling to remove the quartz pinnacles was a good idea as well. D'Eustachio ended his report with an appreciative comment "on the intelligent attitude of the military personnel of the Crystal Branch toward his work."[12]

Though their attitude might have been intelligent, very few were approving of his work. By late October 1943, the Signal Corps leadership had had enough of D'Eustachio's stalling. General Colton decided to temporarily assign both Van Dyke and Bottom to Camp Coles and have them look into the aging study. Their charge was to "investigate the cause of crystal age deterioration, the processes of manufacture which will avoid deterioration," and the "means of establishing inspection techniques for a fundamental quality control program." This is essentially what D'Eustachio had been assigned to do six months earlier.[13] The following month, Major Swinnerton, officer in charge of the QCS, wrote to Camp Coles stating that the aging investigation was "urgent" and that any needed overtime would be approved by his office.[14]

At Camp Coles, Bottom assumed responsibility for the study, essentially carrying out an independent project in parallel with D'Eustachio's. "While I was doing this D'Eustachio sulked and did what he could to impede my efforts," Bottom remembered. Still, he carried on with his work. His first objective was to confirm the effects of humidity and temperature by building a device that came to be known as the "swamp," a chamber within which he could manipulate the temperature and humidity and test their effects upon crystal oscillators. This device would later be used to test the efficacy of various manufacturing methods in combating aging.[15] Noticing significant differences in the aging characteristics of oscillators from different manufacturers, Bottom also began a study of the manufacturing techniques in use across the industry. As expected, the units from those companies that utilized etching suffered aging effects at a much reduced rate.

It is a standard practice in the sciences that experimental results are confirmed by other research groups before their validity is truly accepted. While Bottom was busy with his work, another Camp Coles physicist, 1st Lieutenant Joseph Lukesh, attempted to reproduce the results of D'Eustachio. He was unsuccessful. Experiments through May of 1944 failed to show the effects that D'Eustachio had claimed.[16] Even his X-ray photographs that were purported to show the post-grinding crystal growth became suspect. Bottom managed to trace at least part of the changes shown in the photographs to a vibration in the floor of the lab where the X-ray machine was located.[17]

Scientists outside of Camp Coles were enlisted in the effort as well. The most senior of these was Dr. C.J. Davisson of Bell Laboratories. Davisson had received the Nobel Prize in 1937 for his work on the wave properties of matter and was considered one of the preeminent experimental physicists of his day. D'Eustachio might have coveted a Nobel Prize, but Davisson already had one; and he could find no evidence to support the aging theory put forth by D'Eustachio.[18] This was essentially the "kiss of death" for the theory. D'Eustachio may have seen this coming, or he might simply have tired of Bottom "pulling rank" on him at Camp Coles, but for whatever reason, he resigned from the laboratory and took a position at Bliley Electronics (Bottom felt that he resigned to keep from being fired).[19] On March 13, 1944, Virgil Bottom was named "civilian in charge" of all work carried out at Camp Coles related to oscillator finishing and the causes and effects of aging and eventually became permanently assigned to the lab.[20] From this point on, research was carried out openly and efficiently and results came quickly.

"IT IS UNFORTUNATE THAT SO MUCH TIME AND MONEY (AND NOT A FEW LIVES) WERE LOST WHILE WE WORKED ON A PROBLEM WHOSE SOLUTION WAS ALREADY KNOWN"

Results had already been coming from Bottom's program. By January, he'd reorganized the crystal shop at Camp Coles and had it turning out high-quality crystal units for his tests.[21] His work was considered extremely important by the OCSigO. In February, the transfer of a scientist from the Aircraft Radio Laboratory resulted in their request for Bottom's services. Chief Signal Officer Harry Ingles replied that, no matter how serious their loss was, it could not "be balanced at the expense of [the] fundamental aging program" being carried out by Bottom at Camp Coles.[22] Bottom's first major report came out on March 3, 1944. In it he laid the responsibility for aging on only two factors: the method of lapping and the amount of water vapor present. A very detailed program of experimentation had been carried out. Through it, Bottom discovered the two-stage nature of aging: "An initial aging which is largely completed in 24 to 48 hours" and a "long term aging which appears to be more nearly independent of the treatment of the surface of the crystal but is dependent on humidity."[23] The short duration of the initial stage was the primary reason that the early methods of aging prevention had been considered successful.

Fundamentally, aging was due to the grinding process. No matter how fine the abrasives, a disorganized layer of quartz remained upon the surface of the plate. Loose bits of quartz removed by the lapping could become stuck within cracks in the surface (similar to having food particles stuck between the teeth). Such material could be thrown free during the oscillation of the plate; this was the process of the initial stage of aging. Once the loose quartz particles were freed from the cracked surface, water could enter and begin to slowly react with the quartz. Ultimately, further cracking and loss of quartz from the surface would occur; the second stage of aging. What needed to be done to prevent both stages of aging was to completely remove the disorganized layer. This could only be done by deep etching. The time spent within the acid would have to be long enough to remove not just the layer of loosely held particles but the cracked layer as well. With no particles to come loose, and no cracks for water to enter, a deeply etched plate would not age. Bottom was able to substantiate all of his claims through very detailed experiments and X-ray photographs. Etching could no longer be considered a time- and labor-saving technique available only to a small minority of crystal manufacturers; etching would have to become the only allowed manner in which crystal oscillators could be produced.[24]

Forty years later, in a letter to Charles Bliley, Bottom would bemoan the fact that "so much time and money (and not a few lives) were lost while we worked on a problem whose solution was already known."[25] It is perhaps wrong to blame anyone for failing to see the benefits of etching to final frequency and imposing the method upon the entire industry. Prior to this time, the Signal Corps specifications were written to require particular outcomes: functioning quartz crystal units. No limitations were placed on production methods. Furthermore, the wiping and brushing techniques *appeared to work*. Only Bottom's careful study discovered the true nature of the aging phenomenon and pointed to the ultimate solution. Once the solution had been reached, however, it needed to be communicated to the industry. This was done through three primary means: meetings with the Industry Advisory Committee, QCS reports, and conferences.

Preliminary results regarding the aging problem were being communicated to the Industry Advisory Committee as early as January 1944. The first reports for distribution outside of the Signal Corps were sent out in March. Reports went out to government agencies (such as the Foreign Economic Administration), to allied governments, and to the crystal grinding teams alerting all to the need to etch to final frequency to prevent aging.[26] In July, a summary report of all of Bottom's findings was distributed to all interested agencies.[27]

Reports were good for preliminary announcements and summaries of results, but face-to-face meetings would be required in order to impress upon all crystal manufacturers the importance of changing to the etching process. Plans were laid during the late spring of 1944 for a conference to held in the summer in which all the latest information regarding crystal oscillator specifications, production, and inspection could be disseminated.

"THE SIGNAL CORPS IS OUT TO BUY NOT JUST QUARTZ CRYSTALS, BUT ETCHED QUARTZ CRYSTALS"

By late spring of 1944, so much had changed with respect to the crystal oscillator situation that major revisions in the procedures carried out by Signal Corps inspectors needed to be considered. A conference was held May 7–8 in Chicago at the headquarters of the Signal Corps Inspection Agency. In attendance at the conference were representatives of the Newark, Chicago, Dayton, Philadelphia, and San Francisco Signal Corps Inspection Zones. Also in attendance on the second day of the meeting were Swinnerton and Van Dyke, representing the OCSigO, Bottom and Leo Balter, representing Camp Coles, and others representing Ft. Monmouth and the Aircraft Radio Laboratory. During the meeting, Bottom brought all present up to date on his work regarding aging and the need to etch to final frequency.[28] It was suggested that a conference be held later in the summer at which representatives from all crystal manufacturers could be educated on the aging crisis along with many other changes to oscillator specifications and inspection procedures.[29]

The conference was set for July 11–12 and would be held at the Stevens Hotel in Chicago; Colonel Lester Harris, Director of the Signal Corps Inspection Agency, would serve as the conference chairman. The schedule of presentations included topics related to the proper cleaning of crystal units, the aging problem and the need to convert all plants to utilize acid etching, quality control and inspection, test equipment, and changes in crystal specifications. Opportunities would also exist for small group discussions regarding the production of particular crystal units.[30]

Colonel Harris opened the conference with a reminder of the many instances so far in which the end of the war was claimed to be near: the defeat of the Axis forces in North Africa, the invasion of Italy, the German collapse at Stalingrad, etc. In each case, however, much work was still left to be done. Harris' goal that morning was to convince the assembled crystal manufacturers that a great deal of work was *still* left to be done. In his view, the purpose of the conference was to explain why all crystals must be etched to final frequency and why Signal Corps inspections were going to stress quality control to a greater degree than ever before. It was his belief that, though the war was not over, "through a cooperative effort which results from complete understanding, the war will be won."[31]

Harris was followed in succession by William Halligan, President of Hallicrafters Radio Corporation, and Paul Galvin of the Galvin Manufacturing Corporation. Each speaker reviewed a little of their own history with respect to the crystal industry and urged each and every person there to go back to their plants and redouble their efforts to produce crystal units in both the quantity *and* quality needed.[32] The final speaker of the introductory session was General Colton. In a very thick Southern drawl, Colton remarked upon the tremendous job performed by the industry in turning out the quantities of oscillators needed so far in the war. However, he stressed, the phase during which quantity was more important than quality had passed. Pointing out that failed communications often meant lost battles (and consequently, lost lives), and that only slight deviations in frequency can result in failed communications, Colton urged everyone present to continue in the "splendid" spirit of teamwork that had existed between the Signal Corps and the industry.[33]

After a short break, the results of the May conference at the Inspection Agency were discussed. Lunch was held in the hotel, after which the real work of educating the manufacturers about the aging crisis would begin. The first step was to review the steps that had been taken to date to control the effects of aging. Hugh Waesche of the Aircraft Radio laboratory, speaking on the issue of proper cleaning of quartz plates, pointed out that, even though such methods as toothbrush scrubbing appeared to work, no such techniques completely eliminated aging. However, even though the assembled representatives would be lectured later in the conference on the need to etch crystals, the cleanliness of oscillator plates was still a very important issue that should not be overlooked or minimized.[34]

Karl Van Dyke followed Waesche with a summary discussion of the aging phenomenon. His talk focused on the two fundamental beliefs concerning quartz that had recently been found to be incorrect: that quartz oscillators maintained stable frequencies, and that quartz was essentially chemically inert. Unfortunately, the first belief had been discovered to be false and, from the research recently carried out at Camp Coles, the reason was that the second belief was also false. In fact, it was the chemical "disintegration" of the surface of quartz oscillators that led to the phenomenon of aging. As the conference attendees would learn from Virgil Bottom, the only way to prevent this disintegration was through etching. There were no other options; for the first time, the Signal Corps specifications were going to require a particular production method. From that point on, the Signal Corps would be "out to buy not just quartz crystals, but *etched* quartz crystals."[35]

Given a short break to digest this information, the attendees reconvened to hear a detailed lecture from Bottom on the physics of the aging phenomena. With the ponderous title of "Studies of the Deterioration of Quartz Crystal Units with Special Reference to the Effects of Temperature and Humidity on the Quartz Plate and Holder—Techniques for Fabricating Stable Crystal Units," Bottom's talk covered the details of oscillator physics and the associated processes of aging. Step by step, Bottom outlined the evidence upon which his theory of aging and his suggested solution were based: the quartz powder found on the surface of aged crystals, the effects of humidity and temperature discerned through the use of his artificial "swamp," the X-ray evidence for the disorganized surface layer resulting from even the most gentle lapping, and the complete lack of frequency aging resulting from deep etching. Graphs, photographs, blackboard derivations and calculations, and tables of numbers were shown that supported his conclusions.[36]

During the question-and-answer period following his talk, Bottom was asked about this apparent deviation from performance-based specifications on the part of the Signal Corps. Bottom pointed out that, in terms of aging, the only performance requirement acceptable would be that crystal units not age. As the only way to *test* this would be to hold onto a plant's crystals for several months to determine whether they aged or not, it was felt that the production technique of etching would have to be formally specified.[37] The following day, Captain E.F. Mitchell, chief of the Camp Coles Crystal Branch, made it very clear to the assembled crystal producers that etching to final frequency would become a part of all future crystal specifications and that the change would occur "as soon as possible."[38]

Thus, the industry now had its marching orders. Crystals would be subjected to stricter levels of inspection and quality control and etching would be the only allowed method of final frequency adjustment. As had been the habit of this "most cooperative of all industries" (in the words of General Colton), a great deal of assistance would be available from the members that had already been utilizing etching. In fact, Leon Faber gave a talk on the first afternoon of the conference describing an automatic etching machine that he had developed.[39] The innovation and creativity upon which the crystal industry had been built continued to see it through the aging crisis and the required responses to it. By January 1945, when the aging requirement officially entered the Signal Corps specifications, the industry was producing nearly two and a half million units per month. However, unlike those produced previously, these units stayed on frequency and kept vibrating, allowing the Signal Corps to continue living up to its motto of "getting the message through."

11

"WITHOUT CRYSTALS, YOU HAVE RADIO; WITH THEM, COMMUNICATIONS"

The accomplishments of the quartz oscillator industry during World War II were truly staggering. An industry that produced only a few hundred thousand units during its entire pre-war history was able to produce almost 55 million units during 1942–1944 and an estimated 71 million through the end of the war. To support this industry, over 10 million pounds of raw quartz were imported from Brazil. The increase in production of the industry was extremely rapid, increasing from an estimate of 100,000 units in 1941, to nearly six million in 1942, to over 20 million in 1943 (see Figure 11.1 for monthly and yearly production totals).[1]

These results did not come cheaply, however, whether measured in economic terms or in terms of human effort. Monetarily, the cost of building, supporting, and purchasing the oscillator units from the industry approached $300 million. This cost could have been as much as twice this if not for the numerous advances in production resulting from Signal Corps research and the application of "Yankee ingenuity" in the crystal plants. A 1947 estimate by the Signal Corps concluded that the $3 million it spent on crystal research and development resulted in a savings of nearly $300 million. One example of these savings was the drop in average price over the course of the war from around $8 per unit to $2.[2]

By the final year of the war, the industry was on a solid footing, having successfully weathered three major crises: the initial building of a mass production industry where nothing like it had ever existed, supplying that

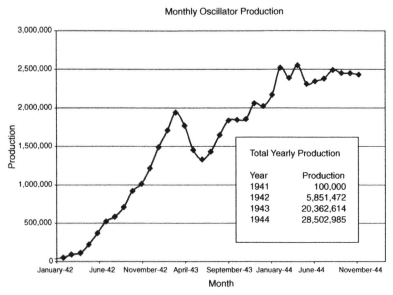

Figure 11.1. Monthly and yearly production totals of quartz crystal units

industry with raw quartz for producing oscillators, and overcoming the aging phenomenon that threatened to cripple the entire industry. Nothing but complete success could have been acceptable, however; not if the troops in the field were going to have the dependable communications technology that they needed to win the war. In the words of James D. O'Connell, "Without crystals, you have radio; with them, communications."[3]

"THE QUARTZ WORKMANSHIP IS VERY MUCH INFERIOR"

The importance of this technology and the industry that supported it cannot be overemphasized. Had the Germans or the Japanese possessed the range and quality of communications equipment that the Allies did, the course of the war might have been different. Though the militaries of both Axis countries did utilize crystal technology, neither came close to the levels of sophistication and dependability that characterized the American products.

The primary method of frequency control utilized by German radios was the master oscillator power amplifier. Periodic calibration was carried out on these radios, with some methods utilizing crystal oscillators. Some radios, such as the Feld fu.b required the use of an identical set for calibration (i.e., the two radios were calibrated to each other). Others, such as the 20 W.S.c, the 100 W.S.c, and the Torn.Fu.g utilized frequency meters for calibrations.

The 30 W.S.a, 80 W.S.a, Torn.Fu.d2, Torn.Fu.b1, and the Kw.E.a used internal crystal-controlled calibration units. For some sets, the radios needed to be operated for several minutes in order to warm up prior to calibration. For others, such as the Torn.Fu.b1, the user's manual warned that calibration of the receiver portion of the set "should not be attempted by operator."[4]

During the war, details of the enemies' crystal-producing capabilities were gleaned from the examination of captured crystals. The capture of enemy equipment was always a goal of combat operations (and so, similarly, was preventing the loss of allied equipment to the enemy; front-line units were instructed never to carry more crystals than were needed to operate their equipment, and radio men were instructed to destroy crystal units in case of imminent capture).[5] Enemy crystal units that were captured ultimately made their way to Camp Coles, where they were studied. Of particular interest early in the war were the methods of weather-proofing used in the crystal units and any information that might be used in developing "counter-measures" against the equipment in which they were used.[6]

In general, the units captured from the Germans and the Japanese were found to be inferior to those being produced in the United States. German crystals were found to exhibit much more variation with temperature than was tolerated in American units. Also, from the errors in orientation observed, it did not appear that the German manufacturers were utilizing X-ray equipment. Some captured units, however, were very impressive from the standpoint of their sealing and weather-proofing.[7]

Due to absolute necessity, however, the German crystal industry made much more efficient use of raw quartz than did either the U.S. or British industries (producing four times as many oscillator plates per pound of raw quartz as the allies). This result was achieved primarily by allowing looser operating tolerances, utilizing predominately low-frequency plates (which were easier to manufacture), and designing the plates to undergo longitudinal vibrations (which meant the plates could be made thinner). The use of low-frequency crystals meant that aging was never really a serious problem for German crystals. Thus, etching to frequency techniques were never developed. Due to the limited supply of raw quartz available, the German crystal manufacturers made use of all quality grades, worrying little about inclusions, needles, or other defects.[8]

Overall, a little less than one million crystal units were estimated to have been produced by the German industry during the war. To a large degree, the German crystal industry was caught as much off guard by the start of the war as were the allied industries.[9] Though some blame can be placed upon the U.S. Signal Corps for the inadequate state of the industry at the time of the Pearl Harbor attack, they had no control over the timing of the war. Germany, on the other hand, *started a war* just as poorly prepared in terms of communications technology and the industry needed to support it. Though Hitler did appear to have a genius for some things, fortunately long-term military planning and strategy weren't among them.

The Japanese utilized frequency control much more so than the German military did. Their oscillator units, however, were of uniformly poor quality. Inspections of captured units found them to be very susceptible to temperature changes and corrosion, to have utilized obsolete designs, and to have been manufactured with inferior quartz workmanship. Furthermore, none examined gave the Signal Corps inspectors any reason to believe they would have passed the operating specifications of the American industry.[10]

One area in which the German crystal program surpassed that of the allies was that of producing synthetic quartz. German industry essentially had to make due with the amount of raw quartz imported up until Brazil cut off all imports to the Axis countries (on the order of 50,000 pounds). This heavily influenced both the design of their crystal units and the amount and nature of crystal control utilized by the German military. Another result was that it stimulated a research program aimed at artificially producing quartz. The leader in this effort was a Professor Nacken of the Mineralogical-Petrographical Institute of the Johannes Wolfgang Goethe University. Professor Nacken had been involved in the synthesis of minerals since 1912. His research during the war focused on producing synthetic quartz from water solutions heated to high temperature and kept under high pressures. For this work, suitable autoclaves were designed and constructed that allowed the researchers to hold their solutions at temperatures around 350°C and pressures of 200–300 atmospheres.[11]

The work was successful. Crushed quartz suspended in the solution was able to precipitate onto seed crystals forming perfectly crystalline blocks. The handedness of the resulting block could be controlled (by controlling the handedness of the seed particle), and untwinned seeds resulted in untwinned blocks. The process was extremely slow, however. A two-centimeter-long crystal could take as long as three weeks to grow. Though the process was successful, and Nacken believed that the techniques could be converted to industrial scales, time ran out on the German Reich before any such industrial-scale program could be established. [12] Thus, an acute shortage of raw quartz placed limitations on the use of crystal control, which placed limits on the effectiveness of communications. Limited communications ultimately played a role in the defeat of the German military. By mid-1943, the American GI was, for the most part, free from the limitations of undependable communications technology. The radio was a weapon, and the American soldier knew how to wield it.

"OUR COMMUNICATIONS HELPED WIN THAT WAR"

One morning near the end of May, 1944, Sergeant Don Alecock was called into his company commander's office. Ordered to draw side arms for himself and a driver, he was informed that he had a special mission. Alecock had been in England for several weeks, assigned to a Signal Company charged with receiving communications equipment from the United States and dispersing it among the units preparing for the invasion of France. Alecock's mission this particular morning was to transport a foot locker–sized crate of quartz crystal units to the Salisbury area. A wholesale change of operating frequencies was planned for the invasion in hopes of confusing the Germans, who had been monitoring the units' radio transmissions for some time.[13] As there was expected to be approximately 90,000 transmitters operating on the land, sea, and in the air of the Normandy invasion areas on D-Day, crystal control was the only way of preventing total chaos on the air.[14] "I have no doubt in my mind," wrote Alecock 55 years later, "that these crystals, changed at the last moment, confused the German communications intelligence and saved many lives on the beaches during the invasion of Europe."[15]

It is safe to say that a great many lives were saved by reliable communications, both before and after the D-Day invasion. Before the Normandy invasion, the war in western Europe was fought primarily in the air. Prior to the war, the Army Air Corps had worked particularly hard at getting its planes equipped with the best radio transmitters and receivers possible, many of which were either frequency controlled or at least calibrated through the use of quartz crystal units. These efforts paid off supremely well in combat. Even with the shortages caused by the aging crisis, it is very difficult to find fliers today who remember any problems with their radios (of course, the Signal Corps worked hard to combat the aging-induced shortages; Virgil Bottom once rushed a crate of crystal units to Bolling Field, where a cargo plane was waiting, engines running, to take it to England).[16] Pilots in particular appear not to have ever given any thought to the reliability of their crystal-controlled command sets. They simply worked when they needed them to. Whether communicating with other pilots, ground controllers, or locating radio beacons, the pilots interviewed for this book found their radios to be extremely reliable. Radio operators, being much more knowledgeable about the technology aboard the planes, have made similar comments: Sidney Rotz, who flew 50 missions in five months with the 15th Air Force in Italy, stated "we never were aware of any crystal crisis." Similarly, Kenneth Lux, flying aboard B-17's and B-24's out of England "was not aware of any shortage of crystals."[17]

Similar reports of the dependability of radio communications come from infantry radio men. In many instances, veterans who did recall problems with communications early in the war pointed out that the situation "improved after crystal equipment arrived."[18] Most, particularly those who entered combat during the final year of the war, remembered no problems at all: "Never had any radio failure in combat" and "I do not remember any particular problems with crystal failure, a shortage of crystals, or any other problems" were common statements.[19] Many of the men who carried the radios across Europe and the Pacific are justifiably proud of the jobs they did. "In the infantry, we had exceptional, often unappreciated, radio communications at the very lowest level; within companies and between companies and battalion," remembered Jack Montrose, a European theater veteran. "It was simply accepted that we had instantaneous contact with each of the companies and with any other special units which might be established. Very seldom did we lose this capability and never because of radio failure." In Montrose's opinion, most of the so-called "radio failures" were primarily the result of poorly trained officers' nonuse or misuse of available radio gear, or bungled intelligence work after the messages had been received. "Some commanders were actually afraid or reluctant to use their radios," fearing German artillery would locate their position through radio direction finding.[20]

Others did not understand the limitations of crystal control. I. Kosmac was a radioman stationed in Iceland. One day his captain entered the radio shack and gave him a message to be transmitted and told him which frequency to transmit it on. "I tried to explain that our crystal (the only one we had) was of a different frequency and that what he wanted couldn't be done. The captain stormed out—a few minutes later the 1st sergeant came in and asked me why I refused to obey the captain's order."[21] By and large, however, field commanders both understood and made the most of their radio equipment. Henry Klingler commanded a Sherman tank across Europe. "Our only means of communication between units was by radio." When asked about the reliability of these radios, Klingler spoke highly of them. "Our radios were quite stable and I was never aware of any major problems with crystals." Perhaps the highest endorsement that could be paid to the radio technology available to the World War II combat soldier was the closing statement of his interview: "Our communications helped win that war."[22]

GAMBLES, SUCCESSES, AND FAILURES

The achievements of the Signal Corps' crystal program were tremendous, particularly since the entire program was a gamble from the very start. To have promoted the adoption of a technology that lacked any means of industrial support could almost be considered a dereliction of duty. Yet, this is exactly what men such as Colton and O'Connell did. The vision of what crystal control could contribute to military communications was so clear in the minds of the Signal Corps research and development men, however, that, in their opinions, they could have done nothing less. Still, they took a huge chance. The crystal manufacturers in existence prior to the war could never have supplied the numbers of crystal units that would have been needed by a peacetime military, much less one actively engaged in war. Any real belief to the contrary on the part of Colton and his cohorts can only be described as having been shortsighted and tremendously optimistic. Even so, under the pressure of war, a supporting industry was created, representing possibly the greatest success of the entire program. The Signal Corps had rolled the dice and, fortunately for the country, the gamble had paid off.

The successful creation of the crystal industry was due to many different factors. First and foremost among them was the existence of a large pool of willing businessmen who had suddenly found themselves either completely out of business or with greatly diminished opportunities. Throughout this group of primarily radio-related entrepreneurs ran a strong streak of "Yankee ingenuity" and a "can do" attitude born of their experiences building businesses during the depression era. Most of these men had already built one business from scratch and it did not seem impossible for them to build another.

The ham radio business had always been characterized by small, one- or two-man operations. The crystal industry built upon this tradition. Whereas other wartime industries, such as the automotive or the aircraft industries, were composed of a few extremely large operations, the crystal industry made use of a large number of these one- or two-man shops. This made it easier for operations to get started during the early months of the war, requiring relatively small investments of capital. On the other hand, small companies can be much more susceptible to the ups and downs of the economy. This collection of small businesses would require a fair amount of external support.

This support came from essentially two sources: through the development of subcontracting networks by the larger radio manufacturers and from the Signal Corps. The collaborative nature of the industry that allowed for smoothly operating subcontracting networks is another result of its roots within the ham radio culture. Hams were known not only for their do-it-yourself creativity, but for their collaboration and sharing of ideas and techniques. The ham radio enterprise had been founded on the concept of communication, after all. Though problems did occur early on due to the proprietary natures of the larger companies, overall, free communication was the order of the day. Whether ideas were shared between individual companies on a face-to-face basis, or with the entire industry through such instruments as the Galvin *Crystal Round Table* and the Signal Corps bulletins, such communication did take place and routinely benefited the entire industry.

The ready-made support structure of the Signal Corps and the civilian government played a large role in the building of the industry as well. Though much work was required to develop the techniques of mass producing crystal oscillators, the methods of supporting such an industry already existed. The Signal Corps maintained staffs of people experienced in the procurement of equipment and the expediting of contracts and the delivery of goods. Furthermore, Depression-era agencies such as the Reconstruction Finance Corporation made the transition to wartime support quickly and smoothly. Construction loans from the RFC and assistance with equipment from the Defense Supplies Corporation enabled businessmen to quickly expand and retool their plants for crystal production.

In addition to the initial lack of companies to produce oscillators was the problem of not knowing how to produce them in mass quantities. This problem was also solved through small-shop ingenuity, Signal Corps support, and intra-industry communication. The research and development laboratories of the Signal Corps had been working on the problems related to mass producing crystal oscillators. As developments occurred, they were communicated to the manufacturers, primarily through the efforts of the Quartz Crystal Section. At the same time, individual manufacturers (in the tradition of the independent ham) were working out their own techniques and designing their own pieces of equipment which were similarly shared with the industry. This combination of top-down and bottom-up development allowed for a greater rate of innovation and implementation.

If any failures can be claimed with regard to the building of the crystal industry, it is that it started too late and was therefore forced to take place too rapidly. Initially consumed with the potential problems of raw quartz supply, little attention was paid by the Signal Corps to the question of actually turning that raw quartz into resonators. It wasn't until a matter of weeks before the Pearl Harbor attack that surveys began to hint at a lack of sufficient production capacity. This, and the country's abrupt entry into the war, led to crisis-mode efforts to recruit and outfit companies for oscillator production. Manufacturers sacrificed quality to produce oscillators in quantity. Problems due to an insufficient level of quality control accounted for a great deal of the early failures of oscillator units in the field. Likewise, the ability to better monitor and examine the crystal units being manufactured likely would have led to an earlier discovery of the aging problem.

The raw quartz procurement effort was another overwhelming success, due primarily to the early start taken by the government and the great degree of cooperation exhibited at the working levels by the various agencies and military departments. The Strategic Materials Act of 1939 supplied both the impetus and the finances for work to begin on the stockpiling of raw quartz. Close cooperation between the Signal Corps and the Treasury Department and the Metals Reserve Corporation got the process off to a quick start. However, failure did occur in one of the vital steps of the procurement process—inspection. The National Bureau of Standards program, under the direction of Dr. Frederick Bates, nearly crippled the entire effort. Far too much time was spent in getting the program underway, in hiring sufficient numbers of inspectors, in building additional facilities financed by the MRC, and establishing an efficient set of inspection specifications.

Perhaps the best response to this deficiency in inspection volume and speed was the decision to station Signal Corps employees in Brazil. This move satisfied several needs. First, it quickly cleared the backlog of raw quartz that had built up in Rio, freeing capital and making way for the purchasing of additional stocks from the interior of the country. Secondly, it gave the U.S. government a means of determining the quality of a shipment before it was purchased and shipped to the National Bureau of Standards. This put the entire program on more of an honest footing with respect to quartz purchases and prices paid. Thirdly, the Americans trained hundreds of Brazilians in the art of quartz inspection and set the standard for how and under what specifications all raw quartz shipments would be evaluated.

Unfortunately, the Signal Corps failed to support these men in a manner befitting their importance to the overall program. Paperwork extending the men's 30–60 day orders was regularly delayed, leaving the more earnest workers wondering about their status and giving the troublemakers support for their malingering and insubordination. Furthermore, the MRC had to step in and cover the travel and per diem expenses for the inspectors when they had to travel as the Signal Corps refused to do so. Even worse, being away from Rio on payday usually meant a delay in receiving their wages from the Army paymaster. Mateson, stationed primarily in Bahia, sometimes went months without receiving his pay. Ultimately, however, the details of supporting the men were worked out and proceeded smoothly for the remainder of the war.

Learning to utilize both smaller and defect-bearing crystals added as much quartz to the stockpile as did the improved inspection procedures worked out by the men in Rio. Coming as it did from the efforts of industry (after yet another Bureau of Standards failure) demonstrates again the ability of the smaller, more flexible companies to respond to a crisis and to affect a solution.

Overcoming the aging crisis represents the most important scientific success of the program. Though a great many breakthroughs in the manufacturing of crystal oscillators occurred, they were primarily of an industrial or an engineering nature. The aging crisis required a careful, experimental, unbiased scientific response. Virgil Bottom's handling of the situation, after weeks of D'Eustachio's secretive work, led to a quick understanding of both the cause and nature of the problem and the proper method of combating it. The wholesale conversion of the industry to the etching process after the July 1944 Chicago conference again points up the flexibility and dedication of the industry.

"PERHAPS THE MOST REMARKABLE OF ALL THE TOOLS SCIENCE HAS GIVEN TO WAR"

So why is it necessary to be writing this book more than 60 years after the end of World War II? Why isn't the story of the quartz crystal oscillator already well known? Hearing the words "scientific contributions to the war effort" usually brings to mind two topics: the atomic bomb and radar. Sometimes the proximity fuse is mentioned in physics textbooks and displayed in museums. The quartz crystal unit, however, is never mentioned. Perhaps the reason that the atomic bomb and radar have been given such places of prominence is that they are primarily associated with a single defining event: for the bomb, the war-ending attacks on Hiroshima and Nagasaki; for radar, the Battle of Britain. Furthermore, the work done on both of these topics is more-or-less associated with particular locations: Los Alamos and the "rad lab" at M.I.T.

The crystal oscillator, on the other hand, did not result from the kinds of emergency crash programs that resulted in the atomic bomb and radar. It was an invention developed in small plants across the country and initially intended primarily for everyday commercial applications and first utilized by amateur radio enthusiasts. At the time, there existed no other reasons for atomic energy or radar than war. The bomb and radar were also much more visible; receiving much more press than did the simple quartz oscillator. Yet, the use of the quartz crystal oscillator was far more common and widespread than even radar. Perhaps it is a true testament to an electronic component's successful operation when its function becomes unseen and taken for granted. This is definitely the case with the quartz crystal unit.

The quartz oscillator has a much more ubiquitous presence and has had a much greater impact on our daily lives than can be claimed for either the atomic bomb or radar. It is true that the specter of nuclear war hung over the world until only the previous few years (though even today the concern about such weapons in the hands of terrorists still exists), and radar contributes to the safe passage of airline travelers around the world. Still, it would be relatively easy to go through a single day without being directly influenced by either. It would be nearly impossible to do so as far as quartz oscillators are concerned. First, we might have to give up any means of timekeeping save for the more "archaic" wind-up watches of two generations or so ago. Cell phone use would be out as well (a near impossibility for many). Long-distance wire communications would also have to be avoided. Color television would be out and, of course, computers. Nearly everything in our lives that requires precise control of frequency or timing utilizes a quartz oscillator. Yet, most people do not even know that they exist. In the 1960s, when society began its move from a mechanical to an electronic world, from an analog to a digital point of view, quartz oscillators were at the forefront. Together with two of the more famous electronic inventions of the previous half century, the transistor and the integrated circuit, quartz oscillators enabled this dramatic shift that ultimately affected all facets of our daily lives.[23]

Just as the wartime oscillator needs of the World War II military could not be met by the one-at-a-time manufacturing processes of the pre-war years, neither could our needs today. Without the innovations that permitted oscillators to be produced at a rate of 2.5 million per month, it is uncertain how long, if ever, it might have taken for these products to become as common as they are today. With transistors and integrated circuits, mass production techniques existed from the very beginning. The fantastic advances these products have enabled could not have occurred if they'd been manufactured by hand, one transistor, or one silicon chip at a time. Invention and production planning went hand in hand. The quartz oscillator did not follow this path. Only through the hard work and extreme dedication of military men, government officials, major industrialists, and basement hobbyists were the techniques invented that allowed the mass production of this most important piece of consumer electronics. Perhaps this saga can be best summed up by a letter written by Professor Gerald Holton of Harvard University to the editors of *Life* magazine following its August 2, 1943 pictorial concerning quartz oscillators:

> In proportion to size, those little glasslike quartz wafers are perhaps the most remarkable of all the tools science has given to war. When the story of the almost incredible progress in research and manufacture of radio crystals in the last two years can be told, it will prove to be a tale of one of this war's greatest achievements. No less significant will be the fruit of these advancements to a new world at peace where crystals will be the vibrating hearts of most telecommunication equipment.[24]

APPENDIX 1

CRYSTAL-CONTROLLED EQUIPMENT

Name	Description	Crystal Type	Crystals
SCR-177	Ground Set	DC-6-A	1
SCR-178	Ground to Air	DC-6	1
SCR-183	Air to Air & Ground	DC-9 or -10	1
SCR-187	Liaison Set for Aircraft	DC-6	1
SCR-188	Ground to Air	DC-6	1
SCR-193	Vehicular Set	DC-6	1
SCR-194	Walkie-Talkie	DC-4	1
SCR-195	Walkie-Talkie	DC-5	1
SCR-197	Ground to Air	FT-171	5
		DC-6	1
SCR-209	Vehicular	DC-6-A	1
SCR-210	Vehicular	DC-6-A	1
SCR-211	Portable Frequency Meter	DC-9	1
SCR-238	Aircraft Set	DC-6-A	1
SCR-240	Aircraft Set	DC-8-A	7
SCR-241	Blind Landing Equip.	DC-13	1
		DC-14	1
SCR-244	Receiver	Special	1
SCR-245	Vehicular	DC-6	1
SCR-253	Remote Control	DC-8	2
SCR-260	Aircraft Set	DC-6-A or -8	1

Crystal Clear: The Struggle for Reliable Communications Technology in World War II, by Richard J. Thompson, Jr.
Copyright © 2007 by Institute of Electrical and Electronics Engineers

SCR-261	Aircraft Set	SC-8-S	7
SCR-268	Radar	Special	1
SCR-270	Radar	Special	1
SCR-271	Radar	Special	1
SCR-274-E	Air to Air and Air to Ground	DC-8-C	2
SCR-274	Command Set	DC-30 or -31	8
SCR-277	Homing Equipment	DC-6-A	1
SCR-281	Radio Telephone Set	FT-249	4
SCR-283	Air to Air and Air to Ground	DC-10-A	1
SCR-284	Vehicular Set	DC-24	1
SCR-287	Aircraft Set	DC-6-A	1
SCR-292	Receiver	IF Special	2
SCR-293	Vehicular Set	FT-171	17
SCR-294	Vehicular Set	FT-171	6
SCR-296	Radar	FT-171	1
SCR-298	Mobile Set	FT-171	3
SCR-299	Mobile, Long Range	DC-6 and FT-171	2
			38
SCR-300	Portable Transceiver	FT-243	2
SCR-506	Vehicular	DC-15 or -24	1
SCR-508	Vehicular	FT-241	80
SCR-509	Vehicular	FT-243	80
SCR-510	Vehicular	FT-243	80
SCR-511	Cavalry Transmitter and Receiver	FT-243	26
SCR-515	UHF Airborne Transmitter and Receiver	FT-243	1
SCR-516	Aircraft Detector	FT-243	1
SCR-518	Radio Altimeter	DC-28	1
SCR-522	VHF Set for Aircraft	DC-11, 16, 26 or CR-1	Varies
SCR-528	Vehicular Command Set	FT-241	80
SCR-533	Radar, IFF	FT-241	1
SCR-536	Paratrooper Transceiver	FT-243	26
SCR-542	VHF Command Set	DC-11 or -16 DC-26 or CR-1	Varies
SCR-543	Anti-Aircraft Set	FT-171	12
SCR-545	Radar	FT-171	1
SCR-549	Television Pick-up Equip.	FT-171	1
SCR-550	Television Pick-up Equip.	FT-171	2
SCR-562	Control Net System	DC-11	32
SCR-563	Control Net System	DC-11	32
SCR-564	Control Net System	DC-11	32
SCR-570	Aircraft Landing Instrument	SC-20	6
SCR-573	Control Net System	DC-11	32

SCR-574	Control Net System	DC-11	32
SCR-583	Cavalry Set	FT-243	8
SCR-585	Receiver-Transmitter	FT-243	6
SCR-591	Aircraft Landing Instrument	DC-8	6
SCR-592	Glide Path Transmitter	Unknown	1
SCR-608	Vehicular Artillery Set	FT-241	120
SCR-609	Artillery Low Power Set	FT-243	120
SCR-610	Artillery Low Power Set	FT-243	120
SCR-611	Compass Locater	DC-8	2
SCR-618	Radio Altimeter	DC-28	1
SCR-619	Vehicular Set	FT-241	22
SCR-624	VHF Set	DC-11, 16, or CR-1	8
SCR-628	Vehicular Set	FT-241	120
SCR-632	Control Net System	DC-11	32
SCR-633	Control Net System	DC-11	32
SCR-643	Control Net System	DC-11	32
SCR-644	Control Net System	DC-11	32
SCR-645	D/F Station	DC-11 or CR-1	192
SCR-694	Field Radio Set	FT-243	2
SCR-696	Receiver	IF Type	4
SCR-698	Transmitter	IF Type	14
SCR-704	Receiver, Crystal Filter	IF Type	1
SCR-708	Vehicular Radio	DC-9, -18, or -19	2
SCR-709	Radio Set	FT-241 or DC-17	19
SCR-728	Vehicular Radio	DC-17	2
SCR-738	Receiver	DC-17	1
SCR-808	Vehicular Radio	DC-18 or -19	2
SCR-809	Receiver & Transmitter	FT-243	22
SCR-810	Radio Set	FT-243	22
SCR-828	Vehicular Radio	DC-18 or -19	1
SCR-838	Radar Set	DC-18 or -19	1

SOURCES:

May 1, 1943, List of Signal Corps Equipment, Record Group 111, OCSigO, Unclassified Central Decimal Files, File 413.44 Crystals, Boxes 1414–1426; National Archives II, College Park, MD. Documents in Chronological Order.

Signal Corps Radio (SCR) Index. Signal Corps Museum Website, available at www.gordon.army.mil/museum.

APPENDIX 2

CRYSTAL MANUFACTURERS

Aircraft Accessories Corporation	Kansas City, MO
American Jewels Corporation	Attleboro, MA
Apex Industries	Chicago, IL
Bangor Electronics Industries	Bangor, ME
Basset, Rex, Inc.	Ft. Lauderdale, FL
Beaumont Electric Company	Chicago, IL
Bendix Radio	Baltimore, MD
Bibber, Marshall G.	Everett, MA
Bliley Electric Company	Erie, PA
Bodner, Charles J., Inc.	Tuckahowe, NY
Brake, H.P., Inc.	
Breon Laboratories	
Burnette Radio Laboratory	San Diego, CA
Butler Crystal Company	Butler, MO
California Electronics Corp.	
Cambridge Thermionics	Cambridge, MA
Carlisle Crystal Company	Carlisle, PA
Chicago Crystal Products Co.	Chicago, IL
Collins Radio Company	Cedar Rapids, IA
Commercial Crystal Co.	Lancaster, PA
Commercial Equipment Co.	Kansas City, MO
Commercial Radio Equipment Co.	Kansas City, MO

Crystal Clear: The Struggle for Reliable Communications Technology in World War II,
by Richard J. Thompson, Jr.
Copyright © 2007 by Institute of Electrical and Electronics Engineers

Connecticut Telephone & Electric	Meriden, CT
Cromar Manufacturing Co.	Williamsport, PA
Crystal Engineering Company	
Crystal Laboratories, Inc.	Wichita, KS
Crystal Products, Inc.	Kansas City, MO
Crystal Research Laboratories	Hartford, CT
C-W Manufacturing Co.	Los Angeles, CA
Dallons Laboratories	Los Angeles, CA
Daughtee Manufacturing Co.	Chicago, IL
Dearborn Scientific Co.	Chicago, IL
D-X Crystal Corp.	Chicago, IL
Dow, L.A.	Seattle, WA
Eastern Quartz Laboratory, Inc.	Dobbs Ferry, NY
Eidson's	Temple, TX
Electric Appliances, Inc.	Indianapolis, IN
Electrical Products Corp.	Oakland, CA
Electronic Industries	Cedar Rapids, IA
Electronic Research Corp.	Chicago, IL
Elkay Radio Products	Oglesby, IL
Federal Engineering Co.	New York, NY
Federal Telephone & Radio Co.	Newark, NJ
Florida Aircraft Radio Corp.	Ft. Lauderdale, FL
Franklin Transformer Man. Co.	Minneapolis, MN
Franklin, L.W.	
Frequency Measuring Service	Kansas City, MO
General Crystal Corp.	Schenectady, NY
General Electric Company	Schenectady, NY
General Piezo Company	Kansas City, MO
General Quartz Laboratory	Irvington on Hudson, NY
General Radio Company	Cambridge, MA
Gentleman Products Co.	Omaha, NE
Goodall Electric Man. Co.	Ogallala, NE
Harvey Radio Laboratories, Inc.	Cambridge, MA
Harvey-Wells Communications	Southbridge, MA
Hatcher and Fisk Man. Co.	Topeka, KS
Hearing Aid Laboratory	Michigan City, IN
Henry Manufacturing Co.	Los Angeles, CA
Higgins Industries, Inc.	Los Angeles, CA
Hi-Power Crystal Company	Chicago, IL
Hoffman, P.R. Co.	Carlisle, PA
Hollister Crystal Company	Boulder, CO
Hunt, G.C. and Sons	Carlisle, PA
James Knights Company	Sandwich, IL
Kaar Engineering Company	Palo Alto, CA
Katz & Ogush, William B., Inc.	New York, NY

Kemlite Laboratories	Chicago, IL
Keystone Piezo Company	Pittsburgh, PA
Knudson Quartz Laboratory	Chicago, IL
Leuck Electric Company	Lincoln, NE
Majestic Radio & Television Corp.	Chicago, IL
Masters Crystal Company	
McGrew Manufacturing Company	Kansas City, MO
Meck, John, Industries	Plymouth, IN
Merit Manufacturing Company	
Midwest Crystal Company	Kansas City, KS
Miller, August E.	North Bergen, NJ
Monitor Piezo Company	South Pasadena, CA
Monowatt Electric Corp.	Providence, RI
Mosteller, Clyde S.	Dallas, TX
Nash California Crystal Co.	Los Angeles, CA
National Scientific Products Co.	Chicago, IL
Nebel, R.N., Laboratory	Brooklyn, NY
North American Philips Co.	Dobbs Ferry, NY
Northern Radio Company	Seattle, WA
Pacific Radio Crystal Company	San Francisco, CA
Palmer, Henry S.	Williamsport, PA
Pan-Elec.ronics Labs.	Atlanta, GA
Pennsylvania Crystal Company	New Kensington, PA
Peterson Radio Company	Council Bluffs, IA
Philco Corporation	Philadelphia, PA
Piezo Electric Products Co.	Brooklyn Park, MD
Precise Development Co.	Chicago, IL
Precision Piezo Service Co.	Baton Rouge, LA
Precision Crystal Laboratories	La Cresenta, CA
Premier Crystal Laboratories	New York, NY
Quartz Crystal Corp. of America	Providence, RI
Quartz Laboratories, Inc.	Kansas City, MO
Quartz Products of New York	New York, NY
R-9 Crystal Company	Pittsburgh, PA
Rack, H.A.	Metuchen, NJ
Radell Corporation	Indianapolis, IN
Radio Specialty Manufacturing Co.	Portland, OR
Ramsey-Young	New York, NY
Rauland Corporation	Chicago, IL
RCA Manufacturing Company	Camden, NJ
Reeves Sound Labs, Inc.	New York, NY
Ring, Carl E.	Lyndhurst, NJ
Ross Manufacturing Co.	Chicago, IL
San Francisco Radio & Supply Co.	San Francisco, CA
Scientific Radio Products Co.	Council Bluffs, IA

Scientific Radio Service	
Sentry Crystal Company	Portland, OR
Shuron Optical Company	Geneva, NY
Silver City Crystal Company	Meriden, CT
Sipp-Eastwood Corporation	Paterson, NJ
Smith, Melvin, L., Laboratories	Kane, PA
Somerset Laboratories	Lyndhurst, NJ
Standard Coil Products Co.	Chicago, IL
Standard Piezo Company	Carlisle, PA, and Scranton, PA
Tedford Crystal Laboratories	Cincinnati, OH
Telicon Corporation	New York, NY
Tru-Lite Research Laboratories	Indianapolis, IN
Turner Company	Cedar Rapids, IA
Union Piezo Company	Newark, NJ
Universal Television System	Kansas City, MO
Urback Development Co.	New York, NY
Valpey Crystal Company	Holliston, MA
V-Precision Instrument Co.	Elmhurst, NY
V-X Company	
Wallace, William T. Company	Peru, IN
Wenkstern-Halsey	Cedar Rapids, IA
Western Electric Company	Chicago, IL, Clifton, NJ, and Kearney, NJ
White Equipment Company	Indianapolis, IN
Wilcox Electric	Kansas City, MO
Wonderlite Company	West Orange, NJ
Wynne Precision Company	Griffin, GA
Zenith Radio Corporation	Chicago, IL

SOURCES:

Melia, Mary-Louise, 1945, The Quartz Crystal Program of the Signal Corps, 1941–1945, Historical Section, Office of the Chief Signal Officer, War Department, Exhibit C.

SC291 Dec 29, 1942 form letter from Maj. Slaughter to industry with attached mailing list. RG111, OCSigO, Unclassified Central Decimal Files, File 413.44 Crystals, Boxes 1414–1426; National Archives II, College Park, MD. Documents in Chronological Order.

REFERENCES

A great many of the documents referenced in the following are contained within a small number of record groups (RG) at the National Archives facility in College Park, Maryland. The particular record group (or box group within a record group) will be identified by codes. The meanings of the codes are as follows:

SCI—RG111, OCSigO, Unclassified Central Decimal Files, File 413.44 Quartz Crystals, Boxes 1476–1477; National Archives II, College Park, MD. Documents in Chronological Order.

SC—RG111, OCSigO, Unclassified Central Decimal Files, File 413.44 Crystals, Boxes 1414–1426; National Archives II, College Park, MD. Documents in Chronological Order.

PLANCOR—RG111, OCSigO, Unclassified Central Decimal Files, File 413.44 095 ExpPLANCOR, Box 225; National Archives II, College Park, MD.

FEA—RG169, Foreign Economic Administration, Entry 165, Box 982, Folder "Import Program—Quartz Crystal," National Archives II, College Park, MD.

VEB—Collection #148, "Virgil Bottom Collection," National Museum of American History, Smithsonian Institute, Washington, DC.

McM—Collected Papers of Virgil E. Bottom, housed at McMurry University, Abilene, Texas

Crystal Clear: The Struggle for Reliable Communications Technology in World War II, by Richard J. Thompson, Jr.
Copyright © 2007 by Institute of Electrical and Electronics Engineers

INTRODUCTION

1. Irwin Gottlieb, personal correspondence, December 11, 1998
2. Zahl, Harold A., "That Fourth Dimension," *Frequency Technology*, April, 1969, pg 31

CHAPTER 1

1. Terrett, D., *The Signal Corps: The Emergency*, from the series *United States Army in World War II*, Center of Military History, 1956, pg 133
2. Same source, pg 154
3. Dear, I.C.B., and Foot, M.R.D. *The Oxford Companion to World War II*, Oxford University Press, Oxford, 1995, pg 923; Perrett, Bryan, *A History of Blitzkrieg*, 1983 New York: Stein and Day pp 86, 126
4. Terrett, D., *The Signal Corps: The Emergency*, from the series *United States Army in World War II*, Center of Military History, 1956, pg 28
5. Same source, pg 88
6. Same source, pg 30
7. Same source, pg 72
8. Same source, pg 107
9. Same source, pg 116
10. Same source, pg 86
11. Same source, pg 87
12. Same source, pg 152
13. Same source, pg 154
14. Same source, pg 155
15. Frondel, Clifford, 1945, "History of the Quartz Oscillator-Plate Industry, 1941–1944," *The American Mineralogist*, Vol 30, pg 205
16. Bottom, Virgil, *The Theory and Design of Quartz Crystal Units*, McMurry Press, Abilene, TX, 1968, pp 2–3
17. Same source, pg 10
18. Parrish, W., 1945, "Methods and Equipment for Sawing Quartz Crystals," *The American Mineralogist*, Vol 30, pp 371–388
19. Frondel, Clifford, 1945, "Final Frequency Adjustments of Quartz Oscillator Plates," *The American Mineralogist*, Vol 30, pg 416
20. Parrish, W., 1945, "Machine Lapping of Quartz Oscillator Plates," *The American Mineralogist*, Vol 30, pp 392–393
21. Frondel, Clifford, 1945, "Final Frequency Adjustments of Quartz Oscillator Plates," *The American Mineralogist*, Vol 30, pp 417–418
22. Bottom, Virgil, *Introduction to Quartz Crystal Unit Design*, Van Nostrand Reinhold, New York, 1982, pg 102; Fowles, G.R. and Cassiday, G.L., *Analytical Mechanics*, Sixth Edition, Saunders College Publishing, Fort Worth, TX, 1999, pg 90

CHAPTER 2

1. Brown, Patrick, 1996, "The Influence of Amateur Radio on the Development of the Commercial Market for Quartz Piezoelectric Resonators in the United States," in *1996 IEEE International Frequency Control Symposium*, pg 58–59; Bottom, Virgil, 1981, "A History of the Quartz Crystal Industry in the USA," in *Proceeding of the 35th Annual Frequency Control Symposium*, pg 3

2. Brown, Patrick, 1996, "The Influence of Amateur Radio on the Development of the Commercial Market for Quartz Piezoelectric Resonators in the United States," in *1996 IEEE International Frequency Control Symposium*, pg 60; Bottom, Virgil, 1981, "A History of the Quartz Crystal Industry in the USA," in *Proceeding of the 35th Annual Frequency Control Symposium*, pg 4

3. Brown, Patrick, 1996, "The Influence of Amateur Radio on the Development of the Commercial Market for Quartz Piezoelectric Resonators in the United States," in *1996 IEEE International Frequency Control Symposium*, pg 59

4. Same source, pg 60

5. QST 1924 Shaw, H.S., 1924, "Oscillating Crystals," in *QST Magazine*, Vol XI, Number 7, July 1924; The monthly publication of the American Radio Relay League and the International Amateur Radio Union, Hartford, CT, pg 30

6. Bottom, Virgil, 1981, "A History of the Quartz Crystal Industry in the USA," in *Proceeding of the 35th Annual Frequency Control Symposium*, pg 4

7. Bottom, Virgil, 1993, *From Possum Holler to Singapore: The Autobiography of Virgil Eldon Bottom*, unpublished memoir, pg 60; March 3, 1981, Letter to Virgil Bottom from L.W. McCoy (VEB)

8. Galvin Poster, *A Timeline of Motorola History*, 1998, Motorola, Inc., Motorola Museum, Schaumburg, IL

9. Bottom, Virgil, 1981, "A History of the Quartz Crystal Industry in the USA," in *Proceeding of the 35th Annual Frequency Control Symposium*, pg 3

10. Laffan, R., 1942, "Quartz Crystals," *Wall Street Journal*, August 3, 1942

11. Melia, Mary-Louise, 1945, *The Quartz Crystal Program of the Signal Corps, 1941–1945*, Historical Section, Office of the Chief Signal Officer, War Department, pg 9

12. Same source, pg 11

13. April 14, 1941, Memo to Major G.K. Heiss, ANMB from George M. Moffett, OPM (SCI)

14. Melia, Mary-Louise, 1945, *The Quartz Crystal Program of the Signal Corps, 1941–1945*, Historical Section, Office of the Chief Signal Officer, War Department, pg 11

15. June 17, 1940, 5th Ind. Letter to Chief of Air Corps from OCSigO, by Col. Eastman, Signal Corps, Executive (SC)

16. Same source

17. January 11, 1940, Report: Comparison of Master Oscillator and Crystal Control for Field Radio Sets (SC)

18. December 21, 1939, Memo to Chief of Infantry from Col. Edwin Butcher; January 15, 1940, Memo to CSigO from Lt. Col. E.W. Fales, Infantry; March 8–April 15, 1940, R&W sheets regarding Crystals for Cavalry (SC)

19. December 21, 1939, Memo to Chief of Infantry from Col. Edwin Butcher; January 15, 1940, Memo to CSigO from Lt. Col. E.W. Fales, Infantry (SC)

20. March 8–April 15, 1940, R&W sheets regarding Crystals for Cavalry (SC)

21. Undated (after March 15, 1941), unsigned draft of report "Crystals for Radio Set SCR-245" (SC)

22. Late August–early September, 1940, R&W sheet regarding crystal-controlled radios (SC)

23. Same source

24. Terrett, D., *The Signal Corps: The Emergency*, 1956, from the series *United States Army in World War II*, Center of Military History, pg 183

25. Same source, pg 142

26. Same source, pg 139

27. Same source, pg 142

28. Same source, pg 142

29. Same source, pg 184

30. Same source, pg 144

31. Coker, Kathy, R. and Stokes, Carol E., 1995, *A Concise History of the US Army Signal Corps*, Office of the Command Historian, US Army Signal Center, Ft. Gordon, pg 22

32. Terrett, D., *The Signal Corps: The Emergency*, 1956, from the series *United States Army in World War II*, Center of Military History, pg 161

33. Same source, pg 164

34. Same source, pg 145

35. August 13, 1940, Memo from Col. Colton to Maj. David Washburn, OCSigO. Subject: Crystal-Controlled Radio Sets (SC)

36. August 10, 1940, Details of August 8 meeting between Col. Roger Colton and Maj. James O'Connell, Signal Corps, and Western Electric representatives (SC)

37. August 13, 1940, Memo from Col. Colton to Maj. David Washburn, OCSigO. Subject: Crystal-Controlled Radio Sets (SC)

38. January 15, 1940, Report by Roger Colton discussing Comparison of Master Oscillator and Crystal Control for Field Radio Sets (SC)

39. Terrett, D., *The Signal Corps: The Emergency*, 1956, from the series *United States Army in World War II*, Center of Military History, pg 164

40. August 10, 1940, Details of August 8 meeting between Col. Roger Colton and Maj. James O'Connell, Signal Corps, and Western Electric representatives (SC)

41. Late August–early September, 1940, R&W sheet regarding crystal-controlled radios (SC)

42. February 17, 1941, Memo to Director, Signal Corp Labs, Ft. Monmouth, from Lt. Col. Hugh Mitchell, Signal Corps; March 6, 1941, Memo to CSigO from Col. Roger Colton (SC)

43. April 23, 1942, R&W sheet from Capt. Abramovich, Air Radio Division, to General Development regarding Elimination of Crystal Filter from Radio Receivers BC-224-( ) (SC)

44. November 22, 1941, Letter from Zenith Radio Corp. to CSigO (SC)

45. October 22, 1940, Letter to CSigO from Lt. Col. Harry Reichelderfer (SCI)

46. January 6, 1941, Summary of Results Obtained on Service Test for Crystal Equipment of Radio Set SCR-245 (SC)

47. January 20, 1941, Memo to CSigO from Lt. Col. Kelly, President of Armored Force Board (SC)

48. February 3, 1941, Memo to CSigO from HQ Armored Force, Cap. Barnes, Assist. Adj. Gen. (SC)

49. Undated (after March 15, 1941), unsigned draft of report "Crystals for Radio Set SCR-245" (SC)

50. Melia, Mary-Louise, 1945, *The Quartz Crystal Program of the Signal Corps, 1941–1945*, Historical Section, Office of the Chief Signal Officer, War Department, pg 16

CHAPTER 3

1. The Stevens Hotel is now the Chicago Hilton & Towers

2. June 24, 1944, Memo to Col. Downing from Maj. Swinnerton; Program for Crystal Conference, Stevens Hotel, Chicago, July 11–12, 1944 (SC)

3. Transcript of Halligan speech from July 11, 1944, Crystal Conference, Historical Resources Collections, Motorola Museum, Motorola, Inc., Schaumburg, IL

4. Transcript of Colton speech from July 11, 1944, Crystal Conference, Historical Resources Collections, Motorola Museum, Motorola, Inc., Schaumburg, IL

5. Nelson, Donald M. *Arsenal of Democracy*, Da Capo Press, New York, 1973, pg 94

6. Transcript of Galvin speech from July 11, 1944, Crystal Conference, Historical Resources Collections, Motorola Museum, Motorola, Inc., Schaumburg, IL; December 17, 1941, Letter to OCSigO from Valpey Crystals; December 27, 1941, Report on a Survey of Quartz Crystal Manufacturing Facilities for Director, Signal Corps Labs by John Fill and Willard Clark (SC); December 30, 1941, letter to CSigO from Harry S. Palmer (SCI)

7. Terrett, Dulany, *United States Army in World War II, The Signal Corps: The Emergency (to December 1941)*, Center of Military History, Washington, DC, 1956, pg 245

8. October 30, 1941, Letter to G.G. Campbell, Radio Specialty Manufacturing, Co. from Capt. H.W. Zermuehlen, Signal Corps (SCI)

9. November 5, 1941, Letter to Robert Freeman of International Telephone & Radio from Capt. H.W. Zermuehlen (SCI)

10. January 5, 1942, Letter to Harry S. Palmer from Capt. H.W. Zermuehlen; January 1, 1942, Letter from Capt. H.W. Zermuehlen to J.W. Baneker, Vice President, Western Electric Company (SCI)

11. January 5, 1942, Letter to Harry S. Palmer from Capt. H.W. Zermuehlen (SCI)

12. January 1, 1942, letter from Capt. H.W. Zermuehlen to J.W. Baneker, Vice President, Western Electric Company (SCI)

13. November 21, 1941, Letter to P.W. Fawlett, District Engineer, OPM from Capt. H.W. Zermuehlen, OCSigO (SCI)

14. Terrett, Dulany, *United States Army in World War II, The Signal Corps: The Emergency (to December 1941)*, Center of Military History, Washington, DC, 1956, pg 255

15. Same source, pg 99, 271

16. Same source, pg 252

17. Same source, pg 271

18. Same source, pg 272

19. Same source, pg 297

20. February 17, 1942, Minutes of February 13 meeting of Coordination & Equipment Division (SC)

21. February 16, 1942, Draft of a memo by James O'Connell (SC)

22. Melia, Mary-Louise, 1945, *The Quartz Crystal Program of the Signal Corps, 1941–1945*, Historical Section, Office of the Chief Signal Officer, War Department, pg 22

23. March 5, 1942, Memo from Brig. Gen. Roger Colton to "All Concerned" (SC)

24. Melia, Mary-Louise, 1945, *The Quartz Crystal Program of the Signal Corps, 1941–1945*, Historical Section, Office of the Chief Signal Officer, War Department, Exhibit F

25. Melia, Mary-Louise, 1945, *The Quartz Crystal Program of the Signal Corps, 1941–1945*, Historical Section, Office of the Chief Signal Officer, War Department, Exhibit F; April 2, 1942, Memo to Lt. Col. O'Connell from Maj. Olsen, Signal Corps (SC)

26. March 29, 1942, Letter from Lt. Col. O'Connell to Mr. Vincent Guldin, National Cash Register Company; July 27, 1942, Letter from Maj. Olsen to John Trittenbach (SC)

27. March 1, 1943, Letter to Prof. Allyn Swinnerton from Col. Downing (SC)

28. Melia, Mary-Louise, 1945, *The Quartz Crystal Program of the Signal Corps, 1941–1945*, Historical Section, Office of the Chief Signal Officer, War Department, Exhibit F

29. September 24, 1943, Memo to Camp Coles from Maj. Gen. Colton and Maj. Swinnerton (SC)

30. Interview with Richard (Dick) Stoiber, September 10, 1998.

31. July 11, 1942, Letter to Bell Laboratory from Maj. Olsen; April 30, 1942, Letters of introduction for James Bell, William Parrish, and E.N. Kagan written to Meade Brunet, Vice President of RCA. by Lt. Col. James O'Connell; June 22, 1942, Telegram from Maj. Gen. Olmstead to Bendix Radio; Bottom, Virgil, 1993, *From Possum Holler to Singapore: The Autobiography of Virgil Eldon Bottom*, unpublished memoir, pg 64 (SC)

32. June 29, 1942, Letter to Apex Industries from Maj. Olsen (SC)

33. April 17, 1942, Memo from Lt. Miller to Major Kauffman (SC)

34. May 24, 1942, Letter to Bendix Radio from Lt. Col. O'Connell (SC)

35. June 24, 1942, Letter to M.L. Smith Labs from Maj. Olsen; December 29, 1942, Form letter from Maj. Slaughter to Industry and attached mailing list (SC)

36. August 11, 1942, Memo to Lt. Col. O'Connell from Lt. Lloyd (SC)

37. December 3, 1942, Letter from Maj. Slaughter to William Hardy & Sons (SC)

38. May 16, 1942, Memo to Lt. Col. O'Connell from Lt. Miller (SC)

39. September 2, 1942, Letter to RCA from Maj. Olsen (SC)

40. September 21, 1942, Letter from Lt. Col. Maier, Dir. SCGDL to Lt. Col. O'Connell (SC)

41. Frondel, Clifford, 1945, "History of the Quartz Oscillator-Plate Industry, 1941–1944," *The American Mineralogist*, Vol 30, pg 210; Bottom, Virgil, 1981, "A History of the Quartz Crystal Industry in the USA," in *Proceeding of the 35th Annual Frequency Control Symposium*, pg 11

42. September 30, 1942, Letter to Philips Metalix Corp. from Maj. Olsen (SC)

43. September 30, 1942, Letter to R-9 Crystals Co. from Maj. Olsen; October 16, 1942, Letter to Philadelphia Signal Depot from Capt. Johnson regarding QCS posters (SC)

44. April 22, 1942, Memo from Lt. Atlass to Lt. Col. Jervey; April 24, 1942, Memo to Director, Ft. Monmouth Labs from Lt. Col. O'Connell (SC)

45. May 5, 1942, Letter to Pratt & Whitney from Lt. Col. O'Connell; August 9, 1942, Letter to RCA from Maj. Olsen (SC)

46. August 9, 1942, Letter to RCA from Maj. Olsen (SC)

47. June 2, 1942, Letter to Emiloid Company from Maj. Kauffman, Signal Corps (SC)

48. Melia, Mary-Louise, 1945, *The Quartz Crystal Program of the Signal Corps, 1941–1945*, Historical Section, Office of the Chief Signal Officer, War Department, pp 25–27

49. June 21, 1942, Memo from William Parrish to Ft. Monmouth (SC)

50. Terrett, Dulany, *United States Army in World War II, The Signal Corps: The Emergency (to December 1941)*, Center of Military History, Washington, DC, 1956, pp 213–214

51. Same source, pg 230

52. Melia, Mary-Louise, 1945, *The Quartz Crystal Program of the Signal Corps, 1941–1945*, Historical Section, Office of the Chief Signal Officer, War Department, pp 25–26

53. January 22, 1943, Letter to Lexington Signal Depot from Lt. Col. C.J. Patche, Signal Corps (SC)

54. November 12, 1942, Memo to CSigO from commander of Lexington Signal Depot; December 14, 1942, Memo to Lexington Signal Depot from Maj. Slaughter (SC)

55. January 22, 1943, Letter to Lexington Signal Depot from Lt. Col. C.J. Patche, Signal Corps (SC)

56. Melia, Mary-Louise, 1945, *The Quartz Crystal Program of the Signal Corps, 1941–1945*, Historical Section, Office of the Chief Signal Officer, War Department, pp 29–30

57. July 10, 1942, Letter to Frederick Bates, National Bureau of Standards, from Maj. Olsen; August 14, 1942, Letter to R.J. Lund, WPB, from Maj. Olsen;

August 17, 1942, Telegram from Maj. Gen. Olmstead to Ft. Monmouth; October 23, 1942, Letter to Morton Bradley from Capt. Johnson (SC)

58. October 16, 1942, Letter to Philadelphia Signal Depot from Capt. Johnson regarding Samuel Gordon; January 7, 1943, Memo from Maj. Slaughter to Director, Camp Coles Signal Laboratory (SC)

59. June 29, 1942, Telegram from Maj. Gen. Olmstead to F. Gana Rodrigues, Rio de Janeiro, Brazil; September 9, 1942, Letter to Board of Economic Warfare, Air Transport Div. from Maj. Olsen; May 11, 1942, Letter to Thomas Perrott from Commercial Crystal Co.; May 17, 1942, Letter to Commercial Crystal Co. from Lt. Col. O'Connell; August 31, 1942, Letter to Apex Industries from Maj. Olsen; September 16, 1942, Telegram from Maj. Gen. Olmstead to Aircraft Access. Co.; October 28, 1942, Memo to Commanding General, AAF from Brig. Gen. Roger Colton (SC)

60. July 22, 1942, Memo to Signal Supply Service from Maj. Olsen (SC)

61. April 18, 1942, Memo from Lt. Lloyd to OCSigO (SC)

62. April 29, 1941, Reports and memos written by C.B. Hamilton, Treasury Department, to CSigO (SC)

63. Melia, Mary-Louise, 1945, *The Quartz Crystal Program of the Signal Corps, 1941–1945*, Historical Section, Office of the Chief Signal Officer, War Department, pg 43

64. February 9, 1942, Telegram from Col. Farmer to Col. Cuny, OCSigO; February 11, 1942, R&W Sheet to Cuny from Procurement Expediting Section, OCSigO (SCI)

65. April 11, 1942, Letter to J.A. Merquelin of Western Electric Co. from Major N.H. Saunders, Signal Corps; June 15, 1942, R&W sheet from Brig. Gen. Roger Colton to Col. Elder, Procurement Division (SC)

66. November 11, 1941, Letter to Capt. H.W. Zermuehlen from P.W. Fawlett, District Engineer, OPM; November 25, 1941, R&W Sheet to a Mr. Campbell from L.H. Niemann, Procurement Planning Section regarding Quartz Crystals (SCI)

67. February 12, 1942, Letter from T.S. Valpey to CSigO (SC)

68. April 24, 1942, Letter to Lt. Col. James O'Connell from Earl Swanson, VP of Apex Industries (SC)

69. April 17, 1942, memo to Director, Ft. Monmouth from Lt. Col. O'Connell; April 23, 1942, Letter from Keystone Radio Supply Co. to Lt. Col. James O'Connell; April 3, 1942, Letter from W.H. Reitz of Aircraft Communications Equipment Corp. to OCSigO; June 27, 1942, Letter to H.L. Wieland from Maj. Olsen; April 25, 1942, Letter from Louis F. Leuck, Leuck Electric Company, to Col. Roger Colton (with attached letter to C. Ridgely Lee (SC); May 7, 1942, Letter to Lt. H.M. Wagner, Aircraft Radio Laboratory, from Matthew Rosenthal of Gabriel Williams Company (SCI); June 11, 1942, Letter to Lt. Col. O'Connell from Premier Radio Enterprises (SC)

70. April 17, 1942, Memo to Director, Ft. Monmouth from Lt. Col. O'Connell; June 27, 1942, Letter to H.L. Wieland from Maj. Olsen (SC)

71. April 17, 1942, Letter from Lt. Col. O'Connell to W.H. Reitz (SC)

72. June 11, 1942, Letter to Lt. Col. O'Connell from Premier Radio Enterprises; June 30, 1942, Letter from Maj. Olsen to Charles K. Morris & Co of Chicago;

April 24, 1942, Memo from Lt. J.J. Lloyd to Lt. Col. James O'Connell; June 22, 1942, Memo from Lt. Lloyd to Maj. Olsen; July 13, 1942, Memo to Maj. Olsen from Lt. Lloyd regarding Aladdin Radio Industries; July 13, 1942, Memo to Maj. Olsen from Lt. Lloyd regarding QOS Corp.; July 20, 1942, Memo to Maj. Olsen from Lt. Lloyd (SC)

73. Melia, Mary-Louise, 1945, *The Quartz Crystal Program of the Signal Corps, 1941–1945*, Historical Section, Office of the Chief Signal Officer, War Department, pg 31

74. Draft of a speech by Virgil Bottom contained in the personal papers of Virgil E. Bottom, housed at McMurry University

75. April 24, 1942, Memo from Lt. J.J. Lloyd to Lt. Col. James O'Connell; July 13, 1942, Memo to Maj. Olsen from Lt. Lloyd regarding Aladdin Radio Industries; June 30, 1942, Letter from Maj. Olsen to Charles K. Morris & Co. of Chicago (SC)

76. July 13, 1942, Memo to Maj. Olsen from Lt. Lloyd regarding QOS Corp. (SC)

77. July 20, 1942, Memo to Maj. Olsen from Lt. Lloyd (SC)

78. April 20, 1942, Memo from Lt. Lloyd to Lt. Col O'Connell (SC)

79. April 22, 1942, Memo from Lt. Lloyd to Lt. Col. O'Connell (SC)

80. "Quartz Crystals for World War II," Presentation by Frank Brewster, n.d., Historical Resources Collections, Motorola Museum, Motorola, Inc., Schaumburg, IL

81. Transcript of Galvin speech from July 11, 1944, Crystal Conference, Historical Resources Collections, Motorola Museum, Motorola, Inc., Schaumburg, IL

82. "Quartz Crystals for World War II," Presentation by Frank Brewster, n.d., Historical Resources Collections, Motorola Museum, Motorola, Inc., Schaumburg, IL; Transcript of Galvin speech from July 11, 1944, Crystal Conference, Historical Resources Collections, Motorola Museum, Motorola, Inc., Schaumburg, IL

83. "Quartz Crystals for World War II," Presentation by Frank Brewster, n.d., Historical Resources Collections, Motorola Museum, Motorola, Inc., Schaumburg, IL; R&W Sheet spanning December 5, 1941, through February 18, 1942, between R&D Division, Procurement Division, and Procurement Planning Section (SC)

84. "Quartz Crystals for World War II," Presentation by Frank Brewster, n.d., Historical Resources Collections, Motorola Museum, Motorola, Inc., Schaumburg, IL

85. Transcript of Halligan speech from July 11, 1944, Crystal Conference, Historical Resources Collections, Motorola Museum, Motorola, Inc., Schaumburg, IL

86. Brewster, Frank, *My Memories of Motorola*, n.d., Historical Resources Collections, Motorola Museum, Motorola, Inc., Schaumburg, IL, pp 14–15

87. March 23, 1942, Minutes from March 19, 1942, meeting between Signal Corps and crystal manufacturers (SC)

88. Melia, Mary-Louise, 1945, *The Quartz Crystal Program of the Signal Corps, 1941–1945*, Historical Section, Office of the Chief Signal Officer, War Department, pg 32

89. April 21, 1942, Cross Reference Sheet by Lt. Col. O'Connell; May 4, 1942, Letter from Lt. Col. O'Connell to J&R Motor Supply Co., Chicago (SC)

90. February 28, 1942, Letter from Capt. H.W. Zermuehlen, CSigO, to P.V. Galvin, Galvin Manufacturing Company (SCI)

91. April 2, 1942, Memo to Lt. Col. O'Connell from Maj. Olsen, Signal Corps (SC)

92. Transcript of a presentation (date unknown, estimated July 1941) given at Army Industrial College regarding history/purpose of the Reconstruction Finance Corporation in defense-related areas (SCI)

93. June 25, 1940, Public Law No. 664, 76th Congress (SCI)

94. Transcript of a presentation (date unknown, estimated July 1941) given at Army Industrial College regarding history/purpose of the Reconstruction Finance Corporation in defense-related areas (SCI)

95. May 16, 1942, Memo to Lt. Col. O'Connell from Maurice Dreusne; May 27, 1942, Letter from Lt. Col. O'Connell to S.R. Udell, RFC; June 22, 1942, Letter to Federal Reserve Bank of Chicago from Lt. Col. O'Connell (SC)

96. April 22, 1942, Letter from Lt. Col. O'Connell to Ernest Smith of RFC (SC)

97. Melia, Mary-Louise, 1945, *The Quartz Crystal Program of the Signal Corps, 1941–1945,* Historical Section, Office of the Chief Signal Officer, War Department, pg 34; Expansion Progress Reports, Budget Bureau Number 49-RO92, Bliley Manufacturing Corporation, June 1943–June 1945 (PLANCOR)

98. Melia, Mary-Louise, 1945, *The Quartz Crystal Program of the Signal Corps, 1941–1945,* Historical Section, Office of the Chief Signal Officer, War Department, pg 34

99. June 24, 1942, Letter to WPB from Maj. Olsen (SC)

100. Same source

101. June 25, 1942, Letter to RFC from Maj. Olsen (SC)

102. July 4, 1942, Letter to Myron Whitney, WPB, from Major Olsen (SC)

103. September 8, 1942, Letter to Brig. Gen. Roger Colton from WPB (SC)

104. Daniels, Roger, *Prisoners Without Trial,* Hill and Wang, New York, 1993, pp 72, 74, 78–79

105. September 2, 1942, Letter to Elmer Wavering of Galvin Manufacturing Corp. from Maj. Olsen; September 6, 1942, Letter to Aircraft Accessories Corp. from Maj. Olsen (SC)

106. October 8, 1942, Letter to Harvard Square Draft Board from Olsen; October 9, 1940, Memo to Army Dispensary from Olsen (SC)

107. April 11, 1981, Letter to Virgil Bottom from Clifford Frondell (VEB); December 31, 1942, Letter from Col. Rives to Col. Marriner (SC)

108. April 11, 1981, Letter to Virgil Bottom from Clifford Frondell (VEB); October 28, 1943, Letter to W.P. Mason, Bell Labs, from Karl Van Dyke (SC)

109. RG111, OCSigO, Unclassified Central Decimal Files, File 413.44 Crystals, Boxes 1414–1426; National Archives II, College Park, MD; Documents in Chronological Order; *passim*

110. May 29, 1943, Letter to Director, General Equipment, RAF from Capt. Johnson; December 16, 1943, Telegram from Maj. Swinnerton to Monitor Piezo; October 23, 1944, Minutes by Maj. Johnson of Crystal Section regarding Crystals for British WS-78 (SC); Quartz Historical Report: Quartz Crystal Section, April 1, 1944, (SCI)

111. April 6, 1943, Letter to Perrott from Capt. Stefan Zamoyski, Polish Assist. Military Attaché; December 27, 1941, Report on A Survey of Quartz Crystal Manufacturing Facilities for Director, Signal Corps Labs by John Fill and Willard Clark; April 22, 1942, Memo from Lt. Col. O'Connell to Director, Ft. Monmouth; January 30, 1943, Letter to Australian War Supply Procurement from Maj. Slaughter (SC)

112. April 13, 1943, Letter to Army-Navy Electronics Production Agency from New Zealand Supply Mission; April 2, 1943, Memorandum for Chief, Materials Branch, Resources and Production Division, HQ, Army Services Forces from Capt. R.G. Wayland; November 5, 1943, Report to OCSigO from Lt. Carfolite, Officer in Charge, Crystal Section, 832 Signal Service Co.; July 2, 1944, Report—Crystal Section Operations, August 25 to June 25, 1944; July 3, 1944, Letter to J.W. Reddrop, RAAF Representative, from Karl Van Dyke (SC)

113. May 15, 1942, Letter from Swedish Legation to War Department; May 19, 1942, Memo from Thomas Perrott to Lt. Col. O'Connell (SC)

114. Dear, I.C.B., and Foot, M.R.D., *The Oxford Companion to World War II*, Oxford University Press, Oxford, 1995, pp 1092–1094

115. May 20, 1942, Memo to Chief, Military Intelligence Service from Lt. Col. O'Connell (SC)

116. June 15, 1942, Letter from Swedish Legation to War Department; June 16, 1942, Memo from Lt. Col. O'Connell to CSigO (SC)

117. November 9, 1942, Memo to Brig. Gen. Roger Colton from Col. O'Connell (SC)

118. June 19, 1942, Memo from Lt. Atlass to Lt. Col. O'Connell; July 4, 1942, Memo to Maj. Stephenson from Adam K. Stricker, Jr.; August 15, 1942, R&W sheet regarding Crystal Requirements (SC)

119. November 9, 1942, Memo to Brig. Gen. Roger Colton from Col. O'Connell (SC)

120. Bottom, Virgil, 1993, *From Possum Holler to Singapore: The Autobiography of Virgil Eldon Bottom*, unpublished memoir, pg 79

121. November 16, 1942, R&W sheet from Col. O'Connell to Col. Rives (SC)

122. November 9, 1942, Memo to Brig. Gen. Roger Colton from Col. O'Connell

123. November 16, 1942, R&W sheet from Col. O'Connell to Col. Rives (SC)

124. February 15, 1943, Memo to International Aid Branch from Lt. J.B. Curtis, Signal Corps (SC)

125. March 23, 1943 (approximate date), Letter to Lt. Col. Messer from Col. Marriner, AAF, Dir. of Communications (SC)

126. March 31, 1943, Letter to Col. Marriner from Lt. Col. Messer (SC)

127. October 14, 1942, Letter from Col. O'Connell to Lt. Col. Messer (SC)

128. Russell, Edward T. and Johnson, Robert M., *Africa to the Alps: The U.S. Army Air Forces in the Mediterranean Theater of Operations*, 1999, Washington, DC, USGPO, pg 122

129. Same source, pg 127

130. Gelb, Norman, *Desperate Venture: The Story of Operation Torch, the Allied Invasion of North Africa*, New York, W. Morrow, pp 223–224

131. Doolittle, James H., *I Could Never Be So Lucky Again*, 1991, New York, Bantam Books, pg 326

132. The Papers of Ira Eaker, Manuscript Division, Library of Congress, Folder I:17 *Correspondence with Arnold, Vol I., December 42–June 43*; December 9, 1942, Letter from Spaatz to Eisenhower

133. The Papers of Ira Eaker, Manuscript Division, Library of Congress, Folder I:17 *Correspondence with Arnold, Vol I., December 42–June 43*; January 2, 1943, Letter from Eaker to Arnold

134. Howe, George F., *US Army in WWII: Mediterranean Theater of Operations: NW Africa: Seizing the Initiative in the West*, 1957, Washington, DC: USGPO, pg 219

135. Same source, pg 280

136. Same source, pg 309

137. Same source, pg 455

138. Same source, pg 380

139. Thompson, G.R., Harris, D.R., Oakes, P.M., and Terrett, D., *United States Army in World War II, The Signal Corps: The Test (December 1941 to July 1943)*, Center of Military History, Washington, DC, 1957, pg 361

140. Howe, George F., *US Army in WWII: Mediterranean Theater of Operations: NW Africa: Seizing the Initiative in the West*, 1957, Washington, DC: USGPO, pg 342

141. Thompson, G.R., Harris, D.R., Oakes, P.M., and Terrett, D., *United States Army in World War II, The Signal Corps: The Test (December 1941 to July 1943)*, Center of Military History, Washington, DC, 1957, pg 335

142. Same source, pg 349

143. Howe, George F., *US Army in WWII: Mediterranean Theater of Operations: NW Africa: Seizing the Initiative in the West*, 1957, Washington, DC: USGPO, pg 161

144. Interview with Roy Smith, June 4, 1999

145. Anderson, Charles R., *US Army in WWII: Guadalcanal, 1st Offensive*, 1949, Washington, DC: USGPO, pg 317

146. Same source, pg 326

147. Same source, pg 317; Fisque, LaVerne interview, June 4, 1999

148. March 23, 1942, Minutes from March 19, 1942, meeting between Signal Corps and crystal manufacturers (SC)

CHAPTER 4

1. February 10, 1981, Letter to Virgil Bottom from Louis Patla (VEB)

2. Fanus, Harriet, "Quartz Crystal Industry in Carlisle, PA," 1983, unpublished

3. Fanus, Harriet "Quartz Crystal Industry in Carlisle, PA," 1983, unpublished; McCommon, Patricia, "A History of the Crystal Industry in the Carlisle Area," History Independent Study, Dickinson College, 1974

4. McCommon, Patricia, "A History of the Crystal Industry in the Carlisle Area," History Independent Study, Dickinson College, 1974

5. April 24, 1943, Letter from C.D. Cuny, Assist. Chief of Signal Supply Services, to L.A. Gagne, Standard Piezo (SC)

6. His plant manager, Luther McCoy, would found yet another Carlisle crystal company, McCoy Electronics, in 1952. March 3, 1981, Letter to Virgil Bottom from L.W. McCoy (VEB)

7. McCommon, Patricia, "A History of the Crystal Industry in the Carlisle Area," History Independent Study, Dickinson College, 1974

8. Bliley, Charles, A. *The Bliley Electric Co., the Early Years: 1930–1955*, Bliley Electric Co. and the Antique Wireless Association, Inc., Holcomb, NY, 1982

9. Same source

10. April 24, 1941, Letter from George Wright, Sales Manager, Bliley Electronics, to Col. Farmer, OCSigO (SCI)

11. December 16, 1980, Letter to Virgil Bottom from John W. Blasier, President, Monitor Products Company (VEB)

12. June 12, 1942, Memo to Lt. Col. O'Connell from Maj. Lippincott regarding Monitor Piezo (SC)

13. Interview with Edith Lineweaver, August 24, 1998

14. June 12, 1942, Memo to Lt. Col. O'Connell from Maj. Lippincott regarding Monitor Piezo (SC)

15. Brown, Patrick, 1996, "The Influence of Amateur Radio on the Development of the Commercial Market for Quartz Piezoelectric Resonators in the United States," in *1996 IEEE International Frequency Control Symposium*, pg 59

16. June 23, 1942, Memo to Lt. Miller from E.K. Woods; September 7, 1942, Letter to Robert McCormick from Maj. Olsen (SC)

17. April 28, 1981, Letter to Virgil Bottom from Herbert Hollister (VEB)

18. Bottom, Virgil, 1993, *From Possum Holler to Singapore: The Autobiography of Virgil Eldon Bottom*, unpublished memoir, pg 51

19. April 28, 1981, Letter to Virgil Bottom from Herbert Hollister (VEB)

20. Same source

21. Bottom, Virgil, 1993, *From Possum Holler to Singapore: The Autobiography of Virgil Eldon Bottom*, unpublished memoir, pg 64

22. Wiseman, Barry, "Scientific Radio Products: The Story of Leo Meyerson's War-Time Crystal Manufacturing," *Electric Radio*, No. 2, June 1989, pp 12–13

23. Same source

24. "Death Ends Career of R.A. Goodall, Nationally-Known Local Industrialist," *World-Herald Newspaper*, Ogallala, NE, October 26, 1953; Bottom, Virgil, 1993, *From Possum Holler to Singapore: The Autobiography of Virgil Eldon Bottom*, unpublished memoir, pg 63

25. "Kansas City's Crystal Pioneers," *Kansas City Star*, May 7, 1974

26. December 27, 1941, Report on A Survey of Quartz Crystal Manufacturing Facilities for Director, Signal Corps Labs by John Fill and Willard Clark (SC)

27. July 11, 1942, Memo to Maj. Olsen from Lt. Lloyd (SC)

28. "Kansas City's Crystal Pioneers," *Kansas City Star*, May 7, 1974 (VEB)

29. Undated (ca. 1980) Letter to Virgil Bottom from Ernest Ruff (VEB)

30. Melia, Mary-Louise, 1945, *The Quartz Crystal Program of the Signal Corps, 1941–1945*, Historical Section, Office of the Chief Signal Officer, War Department, Exhibit C

31. Undated (ca. 1980) Letter to Virgil Bottom from Ernest Ruff (VEB)

32. "Death Ends Career of R.A. Goodall, Nationally-Known Local Industrialist," *World-Herald Newspaper*, Ogallala, NE, October 26, 1953

33. July 17, 1942, Memo to Maj. Olsen from W.E. Richmond (SC)

34. January 19, 1943, Letter from Maj. Slaughter to Kold-Hold Man. Co (SC)

35. February 10, 1981, Letter to Virgil Bottom from Louis Patla (VEB)

36. November 11, 1980, Letter to Virgil Bottom from Leon Faber (VEB)

37. Brewster, Frank, *My Memories of Motorola*, n.d., Historical Resources Collections, Motorola Museum, Motorola, Inc., Schaumburg, IL, pg 28

38. October 25, 1942, Letter to Illinois Northern Utilities Co. from Capt. Johnson; January 11, 1943, Letter to Illinois Northern Utilities Co. from Col. O'Connell (SC)

39. Brewster, Frank, *My Memories of Motorola*, n.d., Historical Resources Collections, Motorola Museum, Motorola, Inc., Schaumburg, IL, pg 28

40. Same source, pp 13–15

41. Same source, pg 17

42. Same source, pg 14

43. Same source, pg 16

44. Parrish, W. and Gordon, S.G. 1945, "Orientation Techniques for the Manufacture of Quartz Oscillator Plates," *The American Mineralogist*, Vol 30, pp 303–325

45. Parrish, W. and Gordon, S.G. 1945, "Precise Angular Control of Quartz Cutting by X-rays," *The American Mineralogist*, Vol 30, pp 326–327

46. Same source, pg 327

47. August 22, 1942, Letter to Hollister Crystal Co from Maj. Olsen; May 30, 1942, Letter from Lt. Col. O'Connell to Franklin Transformer Manufacturing Co.; June 23, 1942, Letter to Lt. Col. O'Connell from Oscar Franck of General Quartz Labs (SC)

48. September 3, 1942, Letter to James Knights from Maj. Olsen (SC)

49. November 11, 1980, Letter to Virgil Bottom from Leon Faber (VEB); Brewster, Frank, *My Memories of Motorola*, n.d., Historical Resources Collections, Motorola Museum, Motorola, Inc., Schaumburg, IL, pg 28

50. Brewster, Frank, *My Memories of Motorola*, n.d., Historical Resources Collections, Motorola Museum, Motorola, Inc., Schaumburg, IL, pg 29

CHAPTER 5

1. Thompson, G.R., Harris, D.R., Oakes, P.M., and Terrett, D., *United States Army in World War II, The Signal Corps: The Test (December 1941 to July 1943)*, Center of Military History, Washington, DC, 1957, pp 542–543

2. Same source, pg 21

3. Melia, Mary-Louise, 1945, *The Quartz Crystal Program of the Signal Corps, 1941-1945*, Historical Section, Office of the Chief Signal Officer, War Department, pg 14

4. March 20, 1942, Notes on Quartz Crystal Production Coordination Meeting by L.H. Niemann, Procurement Planning Section (SCI)

5. September 5, 1942, Telegram from Maj. Gen. Olmstead to Industry members and Signal Corps Depots (SC)

6. Melia, Mary-Louise, 1945, *The Quartz Crystal Program of the Signal Corps, 1941-1945*, Historical Section, Office of the Chief Signal Officer, War Department, pg 39

7. January 27, 1981, Letter to Virgil Bottom from Earl Clark; January 6, 1981, Letter to Virgil Bottom from Jerry Havel (VEB)

8. December 28, 1941, Report on Production of Crystals for Director, Signal Corps Labs by John Hessel; December 27, 1941, Report on A Survey of Quartz Crystal Manufacturing Facilities for Director, Signal Corps Labs by John Fill and Willard Clark (SC)

9. Bliley, Charles A., *The Bliley Electric Co., the Early Years: 1930-1955*, Bliley Electric Co. and the Antique Wireless Association, Inc., Holcomb, NY, 1982

10. December 27, 1941, Report on A Survey of Quartz Crystal Manufacturing Facilities for Director, Signal Corps Labs by John Fill and Willard Clark (SC)

11. December 28, 1941, Report on Production of Crystals for Director, Signal Corps Labs by John Hessel (SC)

12. Bliley, Charles A. *The Bliley Electric Co., the Early Years: 1930-1955*, Bliley Electric Co. and the Antique Wireless Association, Inc., Holcomb, NY, 1982

13. December 27, 1941, Report on A Survey of Quartz Crystal Manufacturing Facilities for Director, Signal Corps Labs by John Fill and Willard Clark (SC)

14. May 5, 1942, Letter to Pratt & Whitney from Lt. Col. O'Connell (SC)

15. February 20, 1942, Minutes of Conference to Discuss Merits of a New Method of Cutting Crystals, and Related Problems (SC)

16. Undated (ca. 1980) Letter to Virgil Bottom from Ernest Ruff (VEB)

17. December 27, 1941, Report on A Survey of Quartz Crystal Manufacturing Facilities for Director, Signal Corps Labs by John Fill and Willard Clark (SC); February 10, 1981, Letter to Virgil Bottom from Louis Patla (VEB)

18. McCommon, Patricia, "A History of the Crystal Industry in the Carlisle Area," History Independent Study, Dickinson College, 1974

19. Parrish, W., 1945, "Machine Lapping of Quartz Oscillator Plates," *The American Mineralogist*, Vol 30, pp 389-415

20. December 26, 1941, Report by Frederick C. Lee, Signal Corps, of an inspection visit to G.C. Hunt & Sons, Carlisle; May 30, 1942, Letter from Lt. Col. O'Connell to Franklin Transformer Manufacturing Co. (SC)

21. June 10, 1942, Memo to Lt. Col. O'Connell from Maj. Olsen regarding Blakeley (SC)

22. Interview with Richard (Dick) Stoiber, September 10, 1998; June 29, 1942, Memo from Richard Stoiber to Lt. Miller; July 31, 1942, Letter to Harvey-Wells from Maj. Olsen; August 26, 1942, Memo to Maj. Olsen from Richard Stoiber (SC)

23. June 11, 1942, Letter from Lt. Col. O'Connell to Donald Sham, WPB; June 11, 1942, Memo to Ft. Monmouth from Lt. Col. O'Connell (SC)

24. June 15, 1942, Memo to Lt. Lloyd from Willie Doxey; June 27, 1942, Memo to Maj. Olsen from Lt. Lloyd; September 21, 1942, Letter to American Instrument Co. from Maj. Olsen (SC)

25. May 30, 1942, Letter from Lt. Col. O'Connell to Franklin Transformer Manufacturing Co. (SC)

26. September 30, 1942, Letter to Philips Metalix Corp. from Maj. Olsen (SC)

27. May 26, 1942, Memo to Director, Ft. Monmouth Labs from Lt. Col. O'Connell; June 15, 1942, Letter to Phillips Metallix Corp. from Lt. Col. O'Connell; June 9, 1942, Results of Meeting Held at General Electric X-ray Corp (SC)

28. October 31, 1942, Memo to Ft. Monmouth from Capt. Johnson (SC)

29. April 10, 1943, Letter to Walter Cady from Col. Downing (SC)

30. April 13, 1942, Memo to CSigO from Lt. Col. Oscar Maier, Signal Corps (SC)

31. Bottom, Virgil, 1993, *From Possum Holler to Singapore: The Autobiography of Virgil Eldon Bottom*, unpublished memoir, pg 59

32. October 31, 1942, Letter to Col. Thompson of Philadelphia SCPD from Col. O'Connell (SC)

33. April 7, 1943, Letter to Lewis Webber, Colorado State College, from Karl Van Dyke (QCS) (SC)

34. Bottom, Virgil, 1993, *From Possum Holler to Singapore: The Autobiography of Virgil Eldon Bottom*, unpublished memoir, pg 60

35. April 7, 1943, Letter to Lewis Webber, Colorado State College, from Karl Van Dyke (QCS); April 7, 1943, letter to Virgil E. Bottom from Karl Van Dyke; April 22, 1943, letter to Virgil E. Bottom from Karl Van Dyke (SC)

36. Bottom, Virgil, 1993, *From Possum Holler to Singapore: The Autobiography of Virgil Eldon Bottom*, unpublished memoir, pg 61

37. Brinkley, David, *Washington Goes to War*, Knopf, New York, 1988, pg 75

38. Bottom, Virgil, 1993, *From Possum Holler to Singapore: The Autobiography of Virgil Eldon Bottom*, unpublished memoir, pp 61–62

39. Same source, pg 64

40. Same source, pg 80

41. June 7, 1942, Memo to HQ, Services of Supply from OCSigO (SC)

42. May 26, 1942, Letter from Lt. Col. O'Connell to Monitor Piezo; September 15, 1942, Letter from Maj. Olsen to the Turner Company, Cedar Rapids, IA; June 26, 1942, Letter to Peterson Radio Company from Maj. Olsen (SC)

43. June 13, 1942, Letter to M.L. Smith Co. from Lt. Col. O'Connell; April 22, 1942, Letter from Lt. Col. James O'Connell to Linwood Gagne of Standard Piezo (SC)

44. June 2, 1942, Letter to Emiloid Company from Maj. Kauffman, Signal Corps; June 20, 1942, Letter to Bronx Draft Board from Lt. Col. O'Connell (SC)

45. August 8, 1942, Letter to Harvey-Wells from Maj. Olsen (SC)

46. June 30, 1942, Letter from Maj. Olsen to Hi Power Crystal Co. (SC)

47. August 5, 1942, Memo from Maj. Olsen to Whom It May Concern regarding "A" Award for A.E. Miller (SC)

48. September 7, 1942, Letter to Robert McCormick from Maj. Olsen (SC)

49. Bottom, Virgil, 1993, *From Possum Holler to Singapore: The Autobiography of Virgil Eldon Bottom*, unpublished memoir, pg 64

50. July 27, 1942, Memo to Commanding Officer, Philadelphia Procurement District, from Lt. Col. Hannah, Signal Corps; September 21, 1942, Telegram from Maj. Gen. Olmstead to Ft. Monmouth (SC)

51. February 16, 1943, Memo to OCSigO from Col. Mack, Signal Corps Contracting Officer; February 23, 1943, Memo from Capt. Johnson to Officer in Charge, Wright Field Signal Corps Procurement District; August 20, 1943, Letter to Smaller War Plants Corp. from Col. Downing (SC)

52. April 27, 1944, Memo from Maj. Swinnerton to Procurement and Distribution Service (SC); Expansion Progress Reports, Budget Bureau Number 49—RO92, Federal Telephone & Radio Corporation, June 1, 1943, August 2, 1943, April 1, 1945, (PLANCOR)

53. Bottom, Virgil, 1993, *From Possum Holler to Singapore: The Autobiography of Virgil Eldon Bottom*, unpublished memoir, pg 67

54. November 30, 1942, Transcript of telephone conversation between Brig. Gen. Roger Colton and Col. Sosthenes Behn (SC)

55. November 29, 1943, Letter to George Field, WPB, from Maj. Swinnerton (SC)

56. Bottom, Virgil, 1993, *From Possum Holler to Singapore: The Autobiography of Virgil Eldon Bottom*, unpublished memoir, pg 68

57. November 29, 1943, Letter to George Field, WPB, from Maj. Swinnerton (SC)

58. April 27, 1944, Memo from Maj. Swinnerton to Procurement and Distribution Service (SC)

59. Virgil Bottom, personal communication; In his memoirs, Bottom refers to the company in question as the Union Electric Company. However, it appears that he has confused this with the Union Piezo Company of Newark, NJ. There was no Union Electric Company of Newark, NJ, under contract to the Signal Corps for crystal production. All of the other details from his account, including the mentioning of Mr. Hawk, agree with the summary of his report on Federal Telephone & Radio contained in the Swinnerton memo (April 27, 1944, Memo from Maj. Swinnerton to Procurement and Distribution Service). Furthermore, the transcript of a telephone conversation between Colton and Behn and a follow-up letter from Behn to Colton seems to confirm the close relationship between them that Bottom alleges. On a subsequent trip to the plant, Bottom was refused entrance—the only time that ever occurred during his entire career in the crystal industry

60. April 2, 1943, Memorandum for Chief, Materials Branch, Resources and Production Division, HQ, Army Services Forces from Capt. R.G. Wayland (SC)

61. April 26, 1943, Memo to Col. O'Connell from Capt. Johnson (SC)

62. April 12, 1943, Letter from P.V. Galvin to Col. E.V. Elder, Director, Materiel Division, OCSigO (SC); Melia, Mary-Louise, 1945, *The Quartz Crystal Program of the Signal Corps, 1941–1945*, Historical Section, Office of the Chief Signal Officer, War Department, pg 42

63. Transcript of Colton speech from July 11, 1944, Crystal Conference, Historical Resources Collections, Motorola Museum, Motorola, Inc., Schaumburg, IL

64. August 19, 1942, Memo to Lt. Col. O'Connell from Samuel Gordon (SC)

65. Same source

66. October 17, 1942, Memo to Col. O'Connell from Wallace Richmond (SC)

67. December 31, 1942, Memos from Carl V. Bertsch (QCS) to Maj. Slaughter (SC)

68. December 31, 1942, Memo to Maj. Slaughter from Carl V. Bertsch (SC)

69. Same source

70. May 18, 1943, Letter from Joseph Egle, Production Engineer of Frequency Measuring Service, Kansas City, MO, to OCSigO (SC)

71. October 23, 1944, Minutes by Maj. Johnson of Crystal Section regarding Crystals for British WS-78 (SC)

72. Bottom, Virgil, 1993, *From Possum Holler to Singapore: The Autobiography of Virgil Eldon Bottom*, unpublished memoir, pg 97

73. Same source, pg 96

74. October 23, 1944, Minutes by Maj. Johnson of Crystal Section regarding Crystals for British WS-78 (SC)

75. Wiseman, Barry, "Scientific Radio Products: The Story of Leo Meyerson's War-Time Crystal Manufacturing," *Electric Radio*, No 2, June 1989, pp 12–13

76. June 10, 1942, Memo to Lt. Col. O'Connell from Maj. Olsen regarding Valinet (SC)

77. September 21, 1942, Letter to Stanley Valinet from Maj. Olsen (SC)

78. Melia, Mary-Louise, 1945, *The Quartz Crystal Program of the Signal Corps, 1941–1945*, Historical Section, Office of the Chief Signal Officer, War Department, Exhibit C

79. RG111, OCSigO, Unclassified Central Decimal Files, File 413.44 Crystals, Boxes 1414–1426; National Archives II, College Park, MD; Documents in Chronological Order; *passim*

80. June 20, 1942, Letter to Bronx Draft Board from Lt. Col. O'Connell; July 23, 1942, Telegram from Maj. Gen. Olmstead to Council Bluffs, IA, draft board; July 23, 1942, Letter to New York Selective Service State HQ from Maj. Olsen (SC)

81. July 23, 1942, Telegram from Maj. Gen. Olmstead to Council Bluffs, IA, draft board (SC)

82. July 6, 1942, Telegram from Maj. Gen. Olmstead to Premier Crystal (SC)

83. July 14, 1942, Letter to Elmer Wavering of Galvin Manufacturing from Maj. Olsen (SC)

84. August 28, 1942, Letter to Standard Radio from Maj. Olsen; July 28, 1942, Letter to A.E. Miller from Maj. Olsen (SC)

85. August 28, 1942, Letter to Standard Radio from Maj. Olsen (SC)

86. July 14, 1942, Letter to Elmer Wavering of Galvin Manufacturing from Maj. Olsen; July 27, 1942, Letter to Galvin Manufacturing Corp. from Maj. Olsen (SC)

87. June 13, 1942, Memo to Officer in Charge, Civilian Personnel from Lt. Col. O'Connell; July 27, 1942, Letter to War Man-Power Commission from Maj.

Olsen; September 6, 1942, Letter to War Manpower Commission from Maj. Olsen (SC)

88. September 6, 1942, Letter to War Manpower Commission from Maj. Olsen (SC)

89. June 23, 1942, Letter to Western Electric from Maj. Olsen (SC)

90. June 13, 1942, Memo to Officer in Charge, Civilian Personnel from Lt. Col. O'Connell (SC)

91. August 2, 1943, Form letter to industry from Lt. Col. Messer (SC)

92. Bottom, Virgil, 1993, *From Possum Holler to Singapore: The Autobiography of Virgil Eldon Bottom*, unpublished memoir, pg 65

93. Memo from Major General H.C. Ingles, CSigO, to War Department Regional Deferment Committee No. 53. Sub: Certification of Leslie H. Balter, April 28, 1944 (Courtesy of Leslie H. Balter)

94. Interview with Hoyt Foster, June 4, 1999

95. Ambrose, Stephen, *Band of Brothers*, 2001, New York, Touchstone, pg 95

96. February 8, 1981, Letter to Virgil Bottom from George Fisher (VEB)

97. Brewster, Frank, *My Memories of Motorola*, n.d., Historical Resources Collections, Motorola Museum, Motorola, Inc., Schaumburg, IL, pg 27

98. Same source, pp 27–28

99. January 18, 1981, Letter to Virgil Bottom from Marvin Bernstein (VEB)

100. Edson, William A., *Vacuum-Tube Oscillators*, John Wiley & Sons, New York, 1953, pp 203–205

101. Melia, Mary-Louise, 1945, *The Quartz Crystal Program of the Signal Corps, 1941–1945*, Historical Section, Office of the Chief Signal Officer, War Department, pp 63–64

102. January 18, 1981, Letter to Virgil Bottom from Marvin Bernstein (VEB)

103. Bottom, Virgil, 1993, *From Possum Holler to Singapore: The Autobiography of Virgil Eldon Bottom*, unpublished memoir, pg 79

104. December 29, 1942, Form letter from Maj. Slaughter to Industry and attached mailing list (SC)

105. January 25, 1943, Form letter to Manufacturers from Maj. Slaughter (SC)

106. Wolfskill, John, "Report on Small Size Crystals," June 3, 1937, and "Small Sized Crystals," August 23, 1940, Bliley Electric Co. Internal Reports, courtesy of Charles Bliley

107. McCormick, R.B., "Quartz Crystal Policies of the War Production Board and Predecessor Agencies, May 1940–August 1945," publisher unknown, pg 96

108. December 12, 1942, Letter to Aircraft Radio Laboratory from Maj. Slaughter (SC)

109. December 24, 1942, letter to JZ Company from Maj. Wood; January 28, 1943, Letter from Frank Cromwell of JZ Co. to Sen. Harry S. Truman; February 3, 1943, Note from Sen. Harry S. Truman to CSigO; February 20, 1943, Letter to Sen. Harry S. Truman from Maj. Gen. Colton (SC)

110. July 31, 1942, Memo to Ft. Monmouth from Maj. Olsen (SC)

111. May 29, 1942, Letter from Lt. Col. O'Connell to Hanslip and Company; July 30, 1942, Letter to W.H. Edwards Company, RI, from Maj. Olsen; October 2, 1942,

Letter from Maj. Olsen to Max Schuster; February 16, 1943, Memo to OCSigO from Col. Mack, Signal Corps Contracting Officer; February 23, 1943, Memo from Capt. Johnson to Officer in Charge, Wright Field Signal Corps Procurement District; August 20, 1943, Letter to Smaller War Plants Corp. from Col. Downing; October 31, 1942, Letter to Col. Thompson of Philadelphia SCPD from Col. O'Connell; December 31, 1942, Memos from Carl V. Bertsch (QCS) to Maj. Slaughter; December 31, 1942, Memo to Maj. Slaughter from Carl V. Bertsch; November 29, 1943, Letter to George Field, WPB, from Maj. Swinnerton; June 23, 1942, Memo to Lt. Miller from E.K. Woods; September 14, 1942, Cover letter for "Handbook for the Manufacture of Quartz Oscillator Plates"; October 7, 1942, Letter to Robert McCormick, WPB, from Maj. Olsen (SC)

112. April 2, 1943, Memorandum for Chief, Materials Branch, Resources and Production Division, HQ, Army Services Forces from Capt. R.G. Wayland (SC); Quartz Historical Report: Quartz Crystal Section, April 1, 1944 (SCI); April 2, 1943, Memorandum for Chief, Materials Branch, Resources and Production Division, HQ, Army Services Forces from Capt. R.G. Wayland (SC); Melia, Mary-Louise, 1945, *The Quartz Crystal Program of the Signal Corps, 1941–1945*, Historical Section, Office of the Chief Signal Officer, War Department, pg 43

CHAPTER 6

1. "Strategic Materials Act, Public, No. 117," June 7, 1939 (S. 572), Library of Congress

2. "A Report to the Congress on Strategic Materials", Submitted by the Army and Navy Munitions Board, November 20, 1944, Library of Congress

3. Strategic Materials Act, Public, No. 117, June 7, 1939 (S. 572), Library of Congress

4. "An Act to Expedite the Strengthening of the National Defense, Public, No. 703," July 2, 1940 (H.R. 9850), Library of Congress

5. Proclamation 2413, Administration of Section 6 of the Act Entitled "An Act to Expedite the Strengthening of the National Defense," July 2, 1940; July 4, 1940, Regulations Governing the Exportation of Articles and Materials Designated in the President's Proclamation of July 2, 1940, Library of Congress

6. February 13, 1940, Letter from Charles Hines, Secretary, ANMB to the RFC containing the approved list of strategic or critical materials, October 7, 1940, Library of Congress

7. Nelson, Donald M., *Arsenal of Democracy*, Da Capo Press, New York, 1973, pg 89

8. McCormick, R.B., "Quartz Crystal Policies of the War Production Board and Predecessor Agencies, May 1940–August 1945," publisher unknown, pg 3

9. August 12, 1940, Minutes of Meeting of the Committee for the Procurement of Quartz Crystal of the National Defense Advisory Commission; August 16, 1940, Memo from Lt. Col. Tom Rives for OCSigO files (SC)

10. August 12, 1940, Minutes of Meeting of the Committee for the Procurement of Quartz Crystal of the National Defense Advisory Commission (SC)

11. August 16, 1940, Memo from Lt. Col. Tom Rives for OCSigO files (SC)

12. September 23, 1940, "Quartz Crystal Procurement Plan" (SCI)

13. September 27, 1940, Memo from Abbott, Radio Engineer, to Maj. Washburn (SC)

14. December 13, 1940, Memo for file from J.E. Gonseth, Jr. (SCI)

15. December 23, 1940, Memo to CSigO from Brig. Gen. H.K. Rutherford, Director, Planning Branch (SCI)

16. February 26, 1941, R&W sheet from Lt. Col. Bogman to Col. Farmer regarding Piezo Crystal (SCI)

17. March 22, 1941, Memo to CSigO from Brig. Gen. H.K. Rutherford, Director, Planning Branch (SCI)

18. Nelson, Donald M., *Arsenal of Democracy,* Da Capo Press, New York, 1973, pg 116

19. Executive Order No. 8629, Establishing the Office of Production Management in the Executive Office of the President and Defining Its Functions and Duties, January 7, 1941, Library of Congress

20. McCormick, R.B., "Quartz Crystal Policies of the War Production Board and Predecessor Agencies, May 1940–August 1945," publisher unknown, pg 1

21. April 14, 1941, Memo to Major G.K. Heiss, ANMB from George M. Moffett, OPM (SCI)

22. Same source

23. April 29, 1941, Reports and memos written by C.B. Hamilton, Treasury Department, to CSigO (SC)

24. McCormick, R.B., "Quartz Crystal Policies of the War Production Board and Predecessor Agencies, May 1940–August 1945," publisher unknown, pg 5

25. Same source, pp 5–6

26. May 22, 1941, Minutes from a meeting at the office of R.J. Lund, (OPM) (SC)

27. Same source

28. Same source

29. McCormick, R.B., "Quartz Crystal Policies of the War Production Board and Predecessor Agencies, May 1940–August 1945," publisher unknown, pg 7

30. Same source, pg 7

31. April 30, 1940, Letter to Secretary of Army & Navy Munitions Board from H. C. Maull, Jr., Director of Treasury Procurement Division (SCI)

32. Same source

33. August 12, 1940, Minutes of Meeting of the Committee for the Procurement of Quartz Crystal of the National Defense Advisory Commission (SC)

34. McCormick, R.B., "Quartz Crystal Policies of the War Production Board and Predecessor Agencies, May 1940–August 1945," publisher unknown, pg 72

35. Briggs, Lyman J., *NBS War Research: The National Bureau of Standards in World War II,* Department of Commerce, 1949, pg 45

36. August 16, 1940, Memo from Lt. Col. Tom Rives for OCSigO files; September 4, 1940, Memo from Lt. Col. Tom Rives to Director, Ft. Monmouth Labs; September 10, 1940, Memo to CSigO from Col. Roger Colton; September 19, 1940,

Memo from Lt. Col. Tom Rives to Director, Ft. Monmouth Labs; September 25, 1940, Memo to CSigO from Col. Roger Colton; September 23, 1940, Memo from Major G.K. Heiss, ANMB, to CSigO (SC)

37. Stoiber, R., Tolman, C. and Butler, R. 1945, "Geology of Quartz Crystal Deposits," *The American Mineralogist*, Vol 30, pg 245–268

38. December 23, 1941, Letter from Bendix Radio to R.J. Lund, OPM (SC); McCormick, R.B., "Quartz Crystal Policies of the War Production Board and Predecessor Agencies, May 1940–August 1945," publisher unknown, pg 64

39. March 18, 1942, Minutes of a meeting on March 11, 1942 (SC)

40. West, F.T. "Quartz Crystal: A Critical and Strategic Mineral," Air University, Maxwell Air Force Base, 1949, pg 14; McCormick, R.B., "Quartz Crystal Policies of the War Production Board and Predecessor Agencies, May 1940–August 1945," publisher unknown, pg 72

41. October 27, 1942, Letter to Administrative Assist. to Secretary of Commerce from Capt. Johnson (SC)

42. McCormick, R.B., "Quartz Crystal Policies of the War Production Board and Predecessor Agencies, May 1940–August 1945," publisher unknown, pg 72

43. March 18, 1942, Minutes of a meeting on March 11, 1942 (SC)

44. McCormick, R.B., "Quartz Crystal Policies of the War Production Board and Predecessor Agencies, May 1940–August 1945," publisher unknown, pg 73

45. Same source, pp 72–73

46. June 29, 1942, Memo to Lt. Col. O'Connell from Maj. Olsen (SC)

47. July 10, 1942, Letter to Frederick Bates, National Bureau of Standards, from Maj. Olsen (SC)

48. August 14, 1942, Letter to R.J. Lund, WPB, from Maj. Olsen (SC)

49. McCormick, R.B., "Quartz Crystal Policies of the War Production Board and Predecessor Agencies, May 1940–August 1945," publisher unknown, pg 74; Briggs, Lyman J., *NBS War Research: The National Bureau of Standards in World War II*, Department of Commerce, 1949, pg 45

50. McCormick, R.B., "Quartz Crystal Policies of the War Production Board and Predecessor Agencies, May 1940–August 1945," publisher unknown, pg 74; October 12, 1942, BEW report regarding new inspection facility at NBS (FEA)

51. McCormick, R.B., "Quartz Crystal Policies of the War Production Board and Predecessor Agencies, May 1940–August 1945," publisher unknown, pg 74

52. August 17, 1942, Telegram from Maj. Gen. Olmstead to Ft. Monmouth; November 5, 1942, Letter to WPB from Capt. Johnson; November 6, 1942, Letter to Purchasing Department Chief of Priorities of NBS from Capt. Johnson; January 12, 1943, Letter to Lyman Briggs, NBS, from Maj. Slaughter (SC)

53. Melia, Mary-Louise, 1945, *The Quartz Crystal Program of the Signal Corps, 1941–1945*, Historical Section, Office of the Chief Signal Officer, War Department, pg 50

54. Briggs, Lyman J., *NBS War Research: The National Bureau of Standards in World War II*, Department of Commerce, 1949, pg 45; McCormick, R.B., "Quartz Crystal Policies of the War Production Board and Predecessor Agencies, May 1940–August 1945," publisher unknown, pg 75

55. McCormick, R.B., "Quartz Crystal Policies of the War Production Board and Predecessor Agencies, May 1940–August 1945," publisher unknown, pp 75–76

56. October 14, 1942, BEW report between Frederick Bates, NBS, and Murray Marker, (FEA)

57. Nelson, Donald M., *Arsenal of Democracy*, Da Capo Press, New York, 1973, pg 194; Executive Order No. 9024, Establishing of the War Production Board in the Executive Office of the President and Defining Its Functions and Duties, January 16, 1942

58. McCormick, R.B., "Quartz Crystal Policies of the War Production Board and Predecessor Agencies, May 1940–August 1945," publisher unknown, pg 2

59. Same source, pg 8

60. Same source, pg 9

61. July 26, 1944, Report "Foreign Economic Administration Country Program for Brazil (Confidential)," RG169, Foreign Economic Administration, Entry 151, Office of Administrator, Records Analysis Div, Historical File 1943–1945, Box 906 "Brazil" Folder, "Country Program Report," National Archives II, College Park, MD

62. November 23, 1942, Quartz Crystal Contract No. 6, Samuel Komisar (FEA)

63. McCormick, R.B., "Quartz Crystal Policies of the War Production Board and Predecessor Agencies, May 1940–August 1945," publisher unknown, pp 28–29

64. September 15, 1941, Letter to Lt. Col Bogman, OCSigO, from D.M. Stoner, Exec. Vice President, Aircraft Accessories Corporation (SCI); May 6, 1942, Letter to Nick Anton, Galvin Manufacturing Corp. from Lt. Col. O'Connell; March 18, 1942, Minutes of March 11, 1942, Conference to Determine Status of Raw Quartz Supply, General Development Section (SC); McCormick, R.B., "Quartz Crystal Policies of the War Production Board and Predecessor Agencies, May 1940–August 1945," publisher unknown, pp 6–7

65. February 27, 1942, Letter from R.J. Lund, WPB, to S.D. Strauss, MRC (SC)

66. June 16, 1942, Letter from Max Schuster to Lt. Col. O'Connell; August 28, 1942, Memo to Maj. Wood from Lt. Col. O'Connell (SC)

67. May 26, 1942, Memo from James Bell, WPB, to R.J. Lund, WPB (SC)

68. Same source

69. McCormick, R.B., "Quartz Crystal Policies of the War Production Board and Predecessor Agencies, May 1940–August 1945," publisher unknown, pg 15

70. September 2, 1942, Letter to Edward Browning, Stockpile and Shipping Branch from S.D. Strauss, Assist. VP MRC (FEA)

71. McCormick, R.B., "Quartz Crystal Policies of the War Production Board and Predecessor Agencies, May 1940–August 1945," publisher unknown, pg 32; September 2, 1942, Letter to Edward Browning, Stockpile and Shipping Branch from S.D. Strauss, Assist. VP MRC (FEA)

72. McCormick, R.B., "Quartz Crystal Policies of the War Production Board and Predecessor Agencies, May 1940–August 1945," publisher unknown, pg 32; West, F.T., "Quartz Crystal: A Critical and Strategic Mineral," Air University, Maxwell Air Force Base, 1949, pg 20

73. McCormick, R.B., "Quartz Crystal Policies of the War Production Board and Predecessor Agencies, May 1940–August 1945," publisher unknown, pg 32

74. Same source, pp 12, 14

75. Same source, pg 35; October 13, 1942, BEW report of meeting between BEW and Ken Murray of Donald Murray Company (FEA)

76. October 13, 1942, BEW report of meeting between BEW and Ken Murray of Donald Murray Company (FEA)

77. McCormick, R.B., "Quartz Crystal Policies of the War Production Board and Predecessor Agencies, May 1940–August 1945," publisher unknown, pg 10

78. Same source, pg 23

79. February 22, 1981, Letter to Virgil Bottom from Willie Doxey (VEB); Doxey, Willie L., 1986, "Quartz Crystals Paved the Way," *Proceedings of the 40th Annual Frequency Control Symposium*, pg 12; February 10, 1981, Letter to Virgil Bottom from Louis Patla (VEB)

80. *Lloyd's War Losses, The Second World War. Vol I: British, Allied, and Neutral Merchant Vessels Sunk or Destroyed by War Causes*, Lloyd's of London, London, Press, 1989

81. Rohwer, Jürgen, Axis Submarine Successes of World War Two: German, Italian, and Japanese Submarine Successes, 1939–1945. Greenhill Books, London, Naval Institute Press, Annapolis, MD, 1999

82. McCormick, R.B., "Quartz Crystal Policies of the War Production Board and Predecessor Agencies, May 1940–August 1945," publisher unknown, pg 25

83. Moore, Arthur, *"A Careless Word . . . A Needless Sinking": A History of the Staggering Losses Suffered by the US Merchant Marine*, American Merchant Marine Museum, King's Point, NY, 1988

84. Rohwer, Jürgen, Axis Submarine Successes of World War Two: German, Italian, and Japanese Submarine Successes, 1939–1945. Greenhill Books, London, Naval Institute Press, Annapolis, MD, 1999

85. July 26, 1944, Report "Foreign Economic Administration Country Program for Brazil (Confidential)," RG169, Foreign Economic Administration, Entry 151, Office of Administrator, Records Analysis Div, Historical File 1943–1945, Box 906 "Brazil" Folder, "Country Program Report," National Archives II, College Park, MD

86. September 9, 1942, Letter to Board of Economic Warfare, Air Transport Div. from Maj. Olsen (SC)

87. Brazilian Bulletin, Brazilian Government Trading Bureau, Vol II, No 26, January 24, 1945, pg 5 (FEA)

88. June 9, 1942, Letter from Lt. Col. O'Connell to Taca Air Lines (SC)

89. September 9, 1942, Letter to Board of Economic Warfare, Air Transport Div. from Maj. Olsen (SC)

90. Interview with Richard Stoiber, September 10, 1998

91. October 16, 1943, Minutes of October 12 meeting. Subject: "Excessive Profits of Importers of Q Xtals" (SC)

92. February 5, 1944, Memo to Director, Purchases Division, from Maj. Gen. Harrison (SC)

93. October 13, 1942, Telegram to Wright Thomas, Liaison Officer, BEW (FEA); June 29, 1942, Telegram from Maj. Gen. Olmstead to F. Gana Rodrigues, Rio de Janeiro, Brazil (SC)

94. October 16, 1943, Minutes of October 12 meeting. Subject: "Excessive Profits of Importers of Q Xtals" (SC)

95. Same source

96. February 27, 1942, Letter from R.J. Lund, WPB, to Kenneth Breon, Breon Radio Lab (SC)

97. May 26, 1942, Memo from James Bell, WPB, to R.J. Lund, WPB (SC)

98. August 31, 1942, Letter to Apex Industries from Maj. Olsen (SC)

99. December 23, 1941, Letter from Bendix Radio to R.J. Lund, OPM (SC)

100. January 6, 1942, Letter from R.J. Lund (OPM) to Bendix Radio (SC)

101. January 3, 1941, Notes from January 2, 1941, meeting of Military and Civilian Advisory Boards (SC)

102. January 17, 1942, Letter from R.J. Lund, OPM, to Harvey-Wells Co.; February 20, 1942, Letter from R.J. Lund, WPB, to A.E. Miller company (SC)

103. May 26, 1942, Memo from James Bell, WPB, to R.J. Lund, WPB (SC)

104. Same source

105. March 9, 1942, Letter from R.J. Lund, WPB, to S.D. Strauss, Vice President of MRC (SC)

106. April 7, 1942, Memo to Lt. Col. James O'Connell from 1st Lt. Charles Miller; April 8, 1942, Memo to Lt. Col. James O'Connell from 1st Lt. Charles Miller; April 10, 1942, Memo to Lt. Col. James O'Connell from 1st Lt. Charles Miller (SC); McCormick, R.B., "Quartz Crystal Policies of the War Production Board and Predecessor Agencies, May 1940–August 1945," publisher unknown, pp 9–10

107. April 7, 1942, Memo to Lt. Col. James O'Connell from 1st Lt. Charles Miller; April 8, 1942, Memo to Lt. Col. James O'Connell from 1st Lt. Charles Miller; April 10, 1942, Memo to Lt. Col. James O'Connell from 1st Lt. Charles Miller; April 20, 1942, Telegram from Major Gen. Olmstead to Maj. Nugent Slaughter, WPB; August 18, 1942, Letter to Robert McCormick, WPB, from Maj. Olsen; August 24, 1942, Note to Robert McCormick from Maj. Olsen (SC)

108. McCormick, R.B., "Quartz Crystal Policies of the War Production Board and Predecessor Agencies, May 1940–August 1945," publisher unknown, pg 49

109. Same source, pg 50

110. Same source, pg 53

CHAPTER 7

1. Stoiber, R., Tolman, C. and Butler, R. 1945, "Geology of Quartz Crystal Deposits," *The American Mineralogist*, Vol 30, pp 254–256

2. Fontes, Oleone Coelho, *Cristais Em Chamas,* Petrópolis, RJ, Vozes, 1993, pg 469

3. Bottom, Virgil, 1993, *From Possum Holler to Singapore: The Autobiography of Virgil Eldon Bottom*, unpublished memoir, pg 68

4. Melia, Mary-Louise, 1945, *The Quartz Crystal Program of the Signal Corps, 1941–1945*, Historical Section, Office of the Chief Signal Officer, War Department, pg 46

5. Stoiber, R., Tolman, C. and Butler, R., 1945, "Geology of Quartz Crystal Deposits," *The American Mineralogist*, Vol 30, pg 253

6. McCormick, R.B., "Quartz Crystal Policies of the War Production Board and Predecessor Agencies, May 1940–August 1945," publisher unknown, pg 27

7. Same source, pg 28

8. Same source, pg 31

9. Same source, pg 34

10. May 3, 1941 Letter to Major C.V. Morgan, Chief, Commodities Division, from Gordon Chambers (SCI)

11. September 7, 1942, Letter to Maj. Wood from Maj. Olsen; September 12, 1942, Memo to Ft. Monmouth from Maj. Olsen (SC)

12. September 24, 1942, Memo to Maj. Wood from Maj. Olsen (SC)

13. September 25, 1942, Memo to U.S. Army Dispensary from Maj. Olsen (SC)

14. October 6, 1942, Telegram to draft board, San Juan, PR, from Maj. Gen. Olmstead; October 10, 1942, Memo to Maj. Wood from Maj. Olsen; October 13, 1942, Letter to San Juan Draft Board from Col. O'Connell; October 15, 1942, Memo to James McKenna, Department of State from Capt. Johnson; October 16, 1942, Memo to Maj. Wood from Capt. Johnson; October 16, 1942, Telegram to Nicholas Colon from Maj. Gen. Olmstead; October 20, 1942, Letter to Paul McGee, USPC, Rio, from Capt. Johnson; October 22, 1942, Letter to Pan Am Airlines from Capt. Johnson (SC); Wynn, J. Clarence and Butler, Robert D., "Report on Signal Corps Personnel Assigned to the Quartz Program, FEA, Brazil, 1 November 1942–31 December 1944," February or March, 1945, pg 5 (FEA)

15. Wynn, J. Clarence and Butler, Robert D., "Report on Signal Corps Personnel Assigned to the Quartz Program, FEA, Brazil, 1 November 1942–31 December 1944," February or March, 1945, pg 5 (FEA)

16. October 14, 1942, Letter from Col. O'Connell to Lt. Col. Messer (SC)

17. October 16, 1942, Letter to Philadelphia Signal Depot from Capt. Johnson regarding Samuel Gordon (SC); Wynn, J. Clarence and Butler, Robert D., "Report on Signal Corps Personnel Assigned to the Quartz Program, FEA, Brazil, 1 November 1942–31 December 1944," February or March, 1945, pg 9 (FEA)

18. Wynn, J. Clarence and Butler, Robert D., "Report on Signal Corps Personnel Assigned to the Quartz Program, FEA, Brazil, 1 November 1942–31 December 1944," February or March, 1945, pg 2, 11 (FEA)

19. Same source, pg 9

20. Same source, pg 22

21. RG 169, Entry 165, Box 982, Folder "Import Program—Quartz Crystal," Memo September 26, 1942, National Archives II, College Park, MD

22. Wynn, J. Clarence and Butler, Robert D., "Report on Signal Corps Personnel Assigned to the Quartz Program, FEA, Brazil, 1 November 1942–31 December 1944," February or March, 1945, pg 11 (FEA)

23. Same source, pg 28

24. February 25, 1943, Letter to James Bell, WPB, from Schmieder, WPB (SC)

25. Wynn, J. Clarence and Butler, Robert D., "Report on Signal Corps Personnel Assigned to the Quartz Program, FEA, Brazil, 1 November 1942–31 December 1944," February or March, 1945, pg 5 (FEA)

26. Same source, pp 11–15

27. Same source, pg 17

28. Same source, pg 18

29. Same source, pg 19

30. Same source, pg 19

31. Interview with Richard (Dick) Stoiber, September 10, 1998

32. April 1, 1943, Memo to Lt. Col. Messer from Maj. Gen. Colton (initialed by Col. O'Connell) (SC)

33. Interview with Richard (Dick) Stoiber, September 10, 1998

34. December 5, 1942, Letter to Maj. Gen. Colton from Arthur Paul, BEW; December 5, 1942, Letter to Maj. Gen. Colton from G. Temple Bridgman, MRC (SC)

35. Same source

36. December 7, 1942, Minutes of Conference on Mining Equipment, Piezo Electric Quartz Crystal, by H.W. Zermuehlen (SCI)

37. December 12, 1942, Letter to Lt. Gen. Somervell from Brig. Gen. Colton; Letter (for signature) from Lt. Gen. Somervell to Donald Nelson, WPB; Spreadsheet listing equipment needed for Brazilian program (SC)

38. Same source

39. December 24, 1942, Letter from Maj. Gen. Colton to Chief, Program Supply Branch, BEW; June 2, 1943, Memo from Lt. Col. Howard, Signal Corps, to Office of the Chief Engineer, International Branch (SC)

40. April 2, 1943, Memorandum for Chief, Materials Branch, Resources and Production Division, HQ, Army Services Forces from Capt. R.G. Wayland; March 11, 1943, Letter to Col. Rives from R.J. Lund, WPB (SC)

41. October 21, 1944, Minutes of "Supply of Raw Quartz in Brazil" by Maj. Johnson; January 14, 1944, Letter to Foreign Economic Administration from Maj. Swinnerton (SC)

42. McCormick, R.B., "Quartz Crystal Policies of the War Production Board and Predecessor Agencies, May 1940–August 1945," publisher unknown, pg 38

43. March 14, 1944, Letter to Alan Bateman, FEA, from Maj. Swinnerton (SC)

44. October 23, 1944, Minutes by Maj. Johnson of Crystal Section regarding Crystals for British WS-78 (SC)

45. June 12, 1944, Memo to Materials Branch, Minerals Section from Maj. Gen. Colton; October 20, 1944, Minutes of "Informal discussion of current quartz situation in Brazil" by Maj. Johnson (SC)

46. October 31, 1944, Minutes of "Informal discussion of quartz situation especially in Brazil" by Maj. Johnson (SC)

47. January 31, 1945, Minutes of "Quartz situation in Brazil" by Maj. Johnson (SC)

48. April 20, 1944, Contract between U.S. Commercial Company and Jose Feliz Bahia, mine owner (FEA)

49. *Congressional Record*, 77th Cong., 1st sess., Feb. 10, 1941, p. 837, Library of Congress

50. Riddle, Donald H., *The Truman Committee: A Study in Congressional Responsibility*, 1964, Rutgers University Press, New Brunswick, NJ, pg 142

51. Report by Senator Hugh Butler to the Special Committee Investigating the National Defense Program and the Joint Committee on Reduction of Nonessential Federal Expenditures. Subject of Report is "Summary of findings in connection with U.S. Expenditures in Latin America"; Letter from Robert Butler in Rio de Janeiro (probably written in December 1943); December 19, 1943, Letter from Mr. Leo Crowley to Senator McKellar (FEA)

52. Same source

53. Though a cooperative atmosphere appears to have existed at the working level, the same cannot be said for the very highest levels of administration. A very bitter feud existed between Vice President Henry Wallace, Chairman of the Board of Economic Warfare, and Jesse Jones, Secretary of Commerce and Chairman of the Reconstruction Finance Corporation. Each blamed the other's organization for impeding the procurement efforts of their own; much of the "dirty laundry" was aired within the pages of the *New York Times* (e.g., *New York Times*, February 16, 1943, pg 5; *New York Times*, July 6, 1943, pg 1; *New York Times*, July 7, 1943, pg 10)

CHAPTER 8

1. McCormick, R.B., "Quartz Crystal Policies of the War Production Board and Predecessor Agencies, May 1940–August 1945," publisher unknown, pg 95

2. Same source, pg 18

3. August 10, 1942, Letter to Byington & Company from Maj. Olsen (SC)

4. October 5, 1942, Letter to Byington & Co. from Maj. Olsen (SC)

5. October 14, 1942, Letter from Col. O'Connell to Lt. Col. Messer (SC)

6. Same source

7. McCormick, R.B., "Quartz Crystal Policies of the War Production Board and Predecessor Agencies, May 1940–August 1945," publisher unknown, pg 45

8. Same source, pg 46

9. Same source, pp 47–48

10. October 14, 1942, Letter to Lt. Col. Philip from Nolan, McNeil & Co.; April 9, 1943, Memo from Capt. Johnson to Ft. Monmouth (SC)

11. McCormick, R.B., "Quartz Crystal Policies of the War Production Board and Predecessor Agencies, May 1940–August 1945," publisher unknown, pg 45

12. Same source, pp 48–49

13. March 4, 1943, Memo to Maj. Gen. Colton from Brig. Gen. Theron Weaver, Resources and Production Division (SC)

14. September 14, 1942, Letter to Elmer Wavering, Galvin Manufacturing Corp., from Maj. Olsen (SC)

15. April 21, 1942, Letter to W.L. Nixon, Industrial Manager, Bausch & Lomb Optical Co. from Lt. Col. James O'Connell; April 25, 1942, letter from W.L. Nixon, Industrial Manager, Bausch & Lomb Optical Co. to Lt. Col. James O'Connell; May 7, 1942, Letter to Director, Signal Corps Labs, from Lt. Col. James O'Connell (SC)

16. April 6, 1943, Letter from Capt. Johnson to Standard Coil (SC)

17. April 17, 1941, Memo to Commodities Division, ANMB, from C.B. Hamilton of Treasury Department; April 21, 1941, "Tentative Specification for Quartz Crystal" by C.B. Hamilton (SCI)

18. September 14, 1942, Letter to Elmer Wavering, Galvin Manufacturing Corp., from Maj. Olsen (SC)

19. February 10, 1981, Letter to Virgil Bottom from Louis Patla (VEB)

20. February 22, 1943, Letter to A. Philip Woolfson and Harold F. Hines, OPM, from Capt. Johnson (SC)

21. March 9, 1943, Memo from Richard Stoiber and Wallace Richmond to Lt. Col. Messer (SC)

22. February 26, 1943, Letter to George Field, WPB, from Lt. Col. Messer (SC)

23. April 17, 1941, Memo to Commodities Division, ANMB, from C.B. Hamilton of Treasury Department; April 21, 1941, "Tentative Specification for Quartz Crystal" by C.B. Hamilton (SCI)

24. September 14, 1942, Report by Lt. Miller for Lt. Col. O'Connell and Maj. Olsen (SC)

25. September 21, 1942, Telegram from Maj. Gen. Olmstead to Ft. Monmouth (SC)

26. March 2, 1943, Letter to Metals Reserve Company from Lt. Col. Messer; September 21, 1942, Telegram from Maj. Gen. Olmstead to Ft. Monmouth (SC)

27. April 26, 1943, Letter from W.L. Clayton, Assist. Sec. Commerce, to Morris Rosenthal, Assist. Dir. BEW (FEA)

28. Briggs, Lyman J., *NBS War Research: The National Bureau of Standards in World War II*, Department of Commerce, 1949, pg 46

29. McCormick, R.B., "Quartz Crystal Policies of the War Production Board and Predecessor Agencies, May 1940–August 1945," publisher unknown, pp 78–79

30. Same source, pp 79–80

31. Same source, pp 81–82

32. Same source, pg 82

33. Same source, pg 83

34. Briggs, Lyman J., *NBS War Research: The National Bureau of Standards in World War II*, Department of Commerce, 1949, pp 45–48

35. McCormick, R.B., "Quartz Crystal Policies of the War Production Board and Predecessor Agencies, May 1940–August 1945," publisher unknown, pg 79

36. January 28, 1940, Letter to Col. Colton from H.C. Dake, Editor, *The Mineralogist* magazine (SC)

37. January 31, 1940, Letter from Col. Bender, Signal Corps, to Director, Ft. Monmouth; February 13, 1940, Letter from Col. Eastman to Director, National

Bureau of Standards; February 21, 1940, Letter from Lyman Briggs, Acting Director, NBS, to Col. Eastman (SC)

38. February 13, March 12, and March 15, 1940, Memos between Navy Yard, Navy Bureau of Engineering, and OCSigO (SC)

39. January 31, 1940, Letter from Col. Bender, Signal Corps, to Director, Ft. Monmouth (SC)

40. April 21, 1942, Letter from Lt. Lloyd to James Bell, WPB (SC)

41. McCormick, R.B., "Quartz Crystal Policies of the War Production Board and Predecessor Agencies, May 1940–August 1945," publisher unknown, pg 58

42. July 17, 1942, Memo to Maj. Olsen from W.E. Richmond (SC)

43. July 11, 1942, Letter to Robert McCormick, WPB, from Maj. Olsen; July 31, 1942, Letter to Donal Hurley from Maj. Olsen (SC)

44. Bobbie J. McLane, e-mail correspondence, February 8, 2002

45. "Montgomery Country—*Our Heritage*," Sesquicentennial Committee, Mount Ida, AR 1986

46. McCormick, R.B., "Quartz Crystal Policies of the War Production Board and Predecessor Agencies, May 1940–August 1945," publisher unknown, pg 58

47. March 1, 1943, Memo from R.J. Lund, WPB, to Howard I. Young, Chairman of Mineral Resources Coordinating Division; April 3, 1943, Letter from Capt. Johnson to W.P. Mason, Bell Laboratories (SC)

48. McCormick, R.B., "Quartz Crystal Policies of the War Production Board and Predecessor Agencies, May 1940–August 1945," publisher unknown, pg 59; April 6, 1943, Letter from Capt. Johnson to Galvin Manufacturing Corporation (SC)

49. Clipping of letter to *Life* magazine, August 23, 1943, responding to an article from August 2, 1943 (VEB)

50. McCormick, R.B., "Quartz Crystal Policies of the War Production Board and Predecessor Agencies, May 1940–August 1945," publisher unknown, pg 59

51. Same source, pg 60

52. January 11, 1943, Letter to U.S. Geological Survey from Maj. Slaughter; March 1, 1943, Memo from R.J. Lund, WPB, to Howard I. Young, Chairman of Mineral Resources Coordinating Division; March 29, 1943, Letter from Leuck Crystal Labs of Lincoln, NE, to Thomas Parrett; July 10, 1943, Letter to a Chas. Johnson of Green River, WY, from Lt. Col. Messer; October 2, 1943, Memo from Maj. Swinnerton to U.S. Dept. of Interior (SC)

53. July 20, 1944, letter from OCSigO to Smaller War Plants Corporation. Author unknown (SC)

54. February 18, 1943, Letters of introduction from Col. John J. Downing, Signal Corps for H.K. Shearer and William T. Pecora; October 5, 1942, Letter to Robert McCormick, WPB, from Maj. Olsen; August 23, 1943, Letter from Capt. Johnson to Professor Snelgrove, Michigan College of Mining and Technology; October 19, 1943, Letter to Frederick Bates, NBS, from Maj. Swinnerton (SC)

55. October 26, 1943, Memo from Maj. Swinnerton to Aircraft Radio Lab (SC)

56. November 27, 1943, Memo to CO, 8th Air Depot, Miami, from Maj. Gen. Colton (SC)

57. December 31, 1943, Letter to Foreign Economic Administration from Maj. Swinnerton (SC)

58. March 19, 1942, Letter from Albert Murray, Communications Section, NDRC, to Lt. Col. O'Connell; April 7, 1942, Letter from Lt. Col. O'Connell to Albert Murray (SC)

59. McCormick, R.B., "Quartz Crystal Policies of the War Production Board and Predecessor Agencies, May 1940–August 1945," publisher unknown, pg 96

60. Same source, pg 97

61. Same source, pg 98

62. Same source

63. Same source, pg 99

64. Same source, pg 100

65. Same source

CHAPTER 9

1. June 1944 Report by Virgil Bottom for Officer in Charge, Crystal Branch, CCSL; April 10, 1944, Memo to SCGSA, Bradley, Beach from Maj. Swinnerton; April 18, 1944, Memo from CCSL to CSigO (SC)

2. Bottom, Virgil, 1993, *From Possum Holler to Singapore: The Autobiography of Virgil Eldon Bottom*, unpublished memoir, pg 71

3. September 3, 1943, Letter to Aircraft Radio Lab from Maj. Gen. Colton; October 20, 1943, Memo to Ft. Monmouth from Col. Downing; November 6, 1943, Telegram from Maj. Swinnerton to Dayton SCPD; December 11, 1943, Memo to Army War College from Brig. Gen. Matejka and Lt. Col. Hodges (SC)

4. July 17, 1944, Memo to CSigO from HQ, CBI, Office of Theater SigO; August 25, 1944, Memo to CSigO from Capt. Hill, Signal Corps (SC)

5. January 8, 1944, Quarterly Historical Report of Quartz Crystal Section (SC)

6. June 25, 1944, Letter to Bendix Radio from J. Walton Colvin, Signal Corps; February 1, 1944, Memo to CSigO from AAF HQ regarding "Reprocessing of Inactive Quartz Crystals" (SC)

7. *Proceedings of the Chicago Crystal Conference, 11–12 July, 1944*, pg 17

8. January 21, 1942, Report to OCSigO from Federal Telegraph; June 29, 1944, Memo to CO, SCGSA from Maj. Swinnerton (SC)

9. Bottom, Virgil, 1993, *From Possum Holler to Singapore: The Autobiography of Virgil Eldon Bottom*, unpublished memoir, pp 66–67

10. September 25, 1943, Report to Commanding Officer, Dayton Signal Corps Procurement District and Depot, from Lt. Col. I.H. Gerks, Chief, C&N Division, Subject: Cleaning of Crystal Oscillator Plates (SC)

11. June 13, 1942, Memo to Lt. Col. O'Connell and Maj. Lippincott from Willie Doxey (SC)

12. May 28, 1942, Memo from Lt. Col. O'Connell to Ft. Monmouth (SC)

13. January 24, 1943, Memo from Col. Rives to Camp Coles (SC)

14. Signal Supply Instructions #95, HQ ETO, July 29, 1944, Subject: Crystal Grinding Teams; courtesy of Mr. Robert Schultz, former grinding team member

15. March 18, 1943, Letter from CSigO to Army Service Forces, Mobilization Branch (SC)

16. July 17, 1943, Letter from Capt. Johnson to Thomas Knoll; July 29, 1943, Letter from Lt. Col. Messer to Mr. Wolfert at Reeves Sound Labs (SC)

17. August 14, 1944, Memo to Miss Selma Greenwald and Miss Vivian Nowicki thru o/c CCSL from Capt. Johnson (SC)

18. Melia, Mary-Louise, 1945, *The Quartz Crystal Program of the Signal Corps, 1941–1945*, Historical Section, Office of the Chief Signal Officer, War Department, pg 84; March 18, 1943, Letter from CSigO to Army Service Forces, Mobilization Branch (SC)

19. May/June 1943, Report of Shipment on Codes 1556 E-F-G-H Sig. 832 Signal Service Co (SC)

20. September 20, 1943, Memo to Chief of Transportation from Maj. Gen. Colton (SC)

21. November 5, 1943, Report to OCSigO from Lt. Carfolite, Officer in Charge, Crystal Section, 832 Signal Service Co.; November 8, 1943, Report to Adjutant General, War Department from Capt. D.E. Wolters, Assistant Officer in Charge, Crystal Grinding Team, 906th Signal Company; January 28, 1944, Report to CSigO from Lt. Kelly, Officer in Charge, Crystal Section, 842nd Signal Service Co. (SC)

22. Thompson, G.R. and Harris, D.R., *United States Army in World War II, The Signal Corps: The Outcome (Mid-1943 through 1945)*, Center of Military History, Washington, DC, 1964, pg 151

23. February 1, 1944, Report to CSigO from Col. Parsons (SC)

24. January 28, 1944, Report to CSigO from Lt. Kelly, Officer in Charge, Crystal Section, 842nd Signal Service Co.; July 2, 1944, Report—Crystal Section Operations, August 25 to June 25, 1944 (SC); February 8, 1981, Letter to Virgil Bottom from George Fisher (VEB)

25. November 5, 1943, Report to OCSigO from Lt. Carfolite, Officer in Charge, Crystal Section, 832nd Signal Service Co. (SC)

26. June 12, 1944, Report to Quartz Crystal Section from Lt. Kelly, Officer in Charge, Crystal Section, 842nd Signal Service Co. (SC)

27. January 29, 1944, Report to CSigO from Lt. Dogan; June 12, 1944, Report to Quartz Crystal Section from Lt. Kelly, Officer in Charge, Crystal Section, 842nd Signal Service Co. (SC)

28. June 12, 1944, Report to Quartz Crystal Section from Lt. Kelly, Officer in Charge, Crystal Section, 842nd Signal Service Co. (SC); February 9, 1981, Letter to Virgil Bottom from H.J. Benedikter (VEB)

29. January 22, 1981, Letter to Virgil Bottom from John D. Holmbeck (VEB); Melia, Mary-Louise, 1945, *The Quartz Crystal Program of the Signal Corps, 1941–1945*, Historical Section, Office of the Chief Signal Officer, War Department, pp 85–86

30. Signal Supply Instructions #95, HQ ETO, July 29, 1944, Subject: Crystal Grinding Teams; courtesy of Mr. Robert Schultz, former grinding team member

31. March 3, 1981, Letter to Virgil Bottom from K.B. Thomson (VEB)

CHAPTER 10

1. Terrett, Dulany, *United States Army in World War II, The Signal Corps: The Emergency (To December 1941)*, Center of Military History, Washington, DC, 1956, pp 27, 33

2. Frondel, Clifford, 1945, "Final Frequency Adjustments of Quartz Oscillator Plates," *The American Mineralogist*, Vol 30, pg 416–460

3. "Aging of Oscillator Plates," Camp Coles internal report, author unknown (McM)

4. Coleman, J., "Confidential Progress Report on TBS Crystal," October 1, 1940 (McM)

5. Same source

6. Coleman, J., "TBS Crystal Investigation," November 29, 1940 (McM)

7. Trouant, V.E, "Confidential Laboratory Report," April 14, 1941 (McM)

8. April 13, 1943, Report written by Van Dyke on "Ageing and Final Finishing Techniques" (SC)

9. September 24, 1943, Memo to Camp Coles from Maj. Gen. Colton and Maj. Swinnerton (SC)

10. Bottom, Virgil, 1993, *From Possum Holler to Singapore: The Autobiography of Virgil Eldon Bottom*, unpublished memoir, pg 71

11. D'Eustachio, D., "Aging of Quartz Crystals," November 8, 1943 (McM)

12. Same source

13. October 27, 1943, Memo from Maj. Swinnerton to Director Engineering & Technical Service; October 29, 1943, Memo to Commanding Officer, SCGSA, Bradley Beach, NJ, from Maj. Gen. Colton (SC)

14. November 20, 1943, Telegram to Camp Coles from Maj. Swinnerton (SC)

15. Bottom, Virgil, 1993, *From Possum Holler to Singapore: The Autobiography of Virgil Eldon Bottom*, unpublished memoir, pg 72

16. May 13, 1944, Report by Lt. Lukesh, subject "Comments on X-ray Data Concerning Surface Changes" (SC)

17. Bottom, Virgil, 1993, *From Possum Holler to Singapore: The Autobiography of Virgil Eldon Bottom*, unpublished memoir, pg 72

18. May 2, 1944, Letter to Lyman Briggs, NBS, from Karl Van Dyke (SC)

19. 1998, Virgil Bottom, personal communication; May 2, 1944. Letter from Karl Van Dyke to C.J. Davisson, Bell Labs (SC)

20. March 13, 1944, Memo to CSigO from CCSL (SC)

21. January 7, 1944, Memo from Col. Downing to Lt. Col. Darke, Development Branch (SC)

22. February 7, 1944, telegram to Director, ARL, Wright Field, from CSigO Harry Ingles (SC)

23. Bottom, V.E., "Crystal Research Memorandum," March 3, 1944 (McM)

24. Same source

25. January 19, 1983, Letter to Charles Bliley from Virgil Bottom (VEB)

26. March 14, 1944, Letter to Alan Bateman, FEA, Maj. Swinnerton; Apr 18, 1944 Memo from Maj. Swinnerton to Commanding Officer, SCGSA; May 5, 1944 Letter to Capt. C.F. Booth from Karl Van Dyke (SC)

27. July 1, 1944, Report by Virgil Bottom entitled "Studies of the Deterioration of QCUs with Special Reference to the Effects of Temperature and Humidity on the Quartz Flats and Holder" (SC)

28. Minutes of Crystal Conference, Headquarters, Signal Corps Inspection Agency, May, 7–8, 1944 (Courtesy of Ken Burch)

29. June 24, 1944, Memo to Col. Downing from Maj. Swinnerton; Program for Crystal Conference, Stevens Hotel, Chicago, July 11–12, 1944 (SC)

30. Same source

31. *Proceedings of the Chicago Crystal Conference, 11—12 July, 1944*, pg 7 (Courtesy of Ken Burch)

32. Same source, pp 9–11

33. Same source, pp 12–14; Sound Recordings from July 11, 1944, Crystal Conference, Historical Resources Collections, Motorola Museum, Motorola, Inc., Schaumburg, IL

34. *Proceedings of the Chicago Crystal Conference, 11–12 July, 1944*, pp 17–21 (Courtesy of Ken Burch)

35. Same source, pp 22–25

36. Same source, pp 26–30

37. Same source, pg 35

38. Same source, pp 65–66

39. Same source, pp 33–34

CHAPTER 11

1. Melia, Mary-Louise, 1945, *The Quartz Crystal Program of the Signal Corps, 1941–1945*, Historical Section, Office of the Chief Signal Officer, War Department, Exhibit J

2. *The Signal Corps Message and Signaleer*, December 17, 1947, pg 3

3. Terrett, D., *The Signal Corps: The Emergency*, from the series *United States Army in World War II*, Center of Military History, 1956, pg 181

4. Barger, Charles, *Radio Equipment of the Third Reich 1933–1945*, pp 28, 32, 40, 61; Barger, Charles, *Communication Equipment of the German Army 1933–1945*, pp 28, 53, 60, 99, 126

5. January 27, 1944, Memo from HQ Fifth Army, subject "Spare Crystals" (SC); Fisque, LaVerne interview, June 4, 1999.

6. September 10, 1943, Memo to Director, Ft. Monmouth, from Maj. Gen. Colton (SC)

7. July 14, 1945, Report by Virgil Bottom on "Foreign Crystals" (SC)

8. Reed, Eugene, D., "German Wartime Crystal Research and Manufacturing Techniques," Division of War Research, Columbia University, New York, date unknown (prior to October 18, 1946) (McM)

9. Same source

10. July 14, 1945, Report by Virgil Bottom on "Foreign Crystals" (SC)

11. Reed, Eugene, D., "German Wartime Crystal Research and Manufacturing Techniques," Division of War Research, Columbia University, New York, date unknown (prior to October 18, 1946) (McM); Reports to Maj. Gen. Van Deusen from R.S. Glasgow, consultant: July 27, 1944, Translated German report on Q synthesis; June 22, 1945, Report to R.S. Glasgow from G.W. Hansell; June 5, 1945, Report by Lt. Col. Slattery (SC)

12. Same sources

13. Alecock, Donald, correspondence, July 18, 1999

14. Thompson, G.R. and Harris, D.R., *United States Army in World War II, The Signal Corps: The Outcome (Mid-1943 through 1945)*, Center of Military History, Washington, DC, 1964, pg 89

15. Alecock, Donald, correspondence, July 18, 1999

16. Bottom, Virgil, 1993, *From Possum Holler to Singapore: The Autobiography of Virgil Eldon Bottom*, unpublished memoir, pg 72

17. Rotz, Sidney, e-mail correspondence, January 25, 1999; Lux, Kenneth, correspondence, April 20, 1999

18. Kalley, Bill, phone interview, June 4, 1999

19. Datthyn, Eugene I., correspondence, May 10, 1999; Macnab, Jim, correspondence, April 23, 1999

20. Montrose, Jack, correspondence, April 29, 1999

21. Kosmac, I., e-mail correspondence, May 19, 1999

22. Klingler, Henry, phone interview, November 14, 1998

23. Stephens, C. and Dennis, M., "Engineering Time: Inventing the Electronic Wristwatch," *BJHS*, 2000, Vol. 33, pp 477–497

24. *Life*, August 23, 1943 (VEB)

INDEX

Crystal Clear: The Struggle for Reliable Communications Technology in World War II,
by Richard J. Thompson, Jr.
Copyright © 2007 by Institute of Electrical and Electronics Engineers

Printed and bound by CPI Group (UK) Ltd, Croydon, CR0 4YY

27/10/2024

14580341-0001